D0824365

NEW INVENTIONS
IN LOW COST SOLAR HEATING

100 Daring Schemes Tried and Untried

NEW INVENTIONS IN LOW COST SOLAR HEATING

100 Daring Schemes Tried and Untried

William A. Shurcliff

BRICK HOUSE PUBLISHING COMPANY
Harrisville, N.H.   Andover, Mass.

Published by: Brick House Publishing Co.

Editorial   3 Main St.
offices: Andover, Mass. 01810

Customer   Church Hill
Service: Harrisville, N.H. 03450

Production credits: Composition and illustrations: Wilkins & Associates
Copy editor: Jim Bright
Cover design: Ned Williams
Editor: Jack Howell

Printed in the United States of America
Library of Congress Catalog Card Number: 79-50275
ISBN: 0-931790-02-6

Copyright © 1979 by William A. Shurcliff.

All rights reserved. No part of this book may be reproduced in any form without the
written permission of the publisher.

Contents

viii

Part 3   **ACTIVE SYSTEMS [NON-CONCENTRATING]**

X

xii

Introduction

NEED FOR CHEAPER SOLAR HEATING SYSTEMS

Most kinds of solar heating systems cost too much. Nearly all solar architects and engineers know this; they don't like to talk about it, but they have it on their minds continually. They know, for example, that an active–type solar heating system for a typical house in a cold region in USA costs, typically, $7000 to $15,000, and the first-year saving of fuel oil may amount to only about 1/20 of this.

Passive systems cost less. Typically, they cost about half as much as active systems. Maintenance and operating costs tend to be much lower also. But many houses having passive solar heating systems tend to be too hot on sunny days and too cold on overcast days. The amount of heat stored in massive floors and walls is small, unless the occupants are willing to accept a large swing in indoor temperature, e.g., a swing of 10 or 20 F degrees.

Hundreds of engineers, architects, and inventors have been trying for years to invent solar heating systems that will cost less. Much progress has been made. Wide publicity has been given to systems developed by H. R. Hay, S. Baer, H. E. Thomason, N. B. Saunders, B. Anderson, and D. Wright, to name but a few. Many of their solar buildings are described in my 1977 book, *Solar Heated Buildings: A Brief Survey*, 13th ed., and in my 1978 book, *Solar Heated Buildings of North America: 120 Outstanding Examples*, published by Brick House Publishing Co., Harrisville, NH 03450; $8.95. See also important books by Bruce Anderson, D. R. Watson, the American Institute of Architects, DOE, HUD, and others.

But not enough has been done. The need for systems that are less expensive, and for inexpensive systems that perform better, is still acute. In the last six years I have tried hard to invent cheaper systems. Some of my inventions are entirely new, while others are mere variations or improvements on other inventors' ideas. Usually I have made a careful write-up of each such invention or improvement. Many of these write-ups, edited and shortened, are presented here. Some accounts of unusual inventions by others are included also.

KINDS OF INVENTIONS INCLUDED

The inventions described here deal with low-cost solar heating systems. They deal with passive systems, active systems, and combination passive-and-active systems, including some instantly convertible passive-to-active systems. Of the active systems described, some employ air and some employ water. Some employ flat-plate collectors and some employ concentrating collectors. Some of the inventions are concerned just with thermal storage systems: water-filled tanks, bins-of-stones, bins-of-water-filled bottles, and containers filled with phase-change materials such as Glauber's salt or hypo.

Most of the solar heating systems described are for space heating. A few are for domestic hot water.

WHOM IS THE BOOK MEANT FOR?

For inventors, architects, and designers of solar heated buildings. Also for those engineers and builders who recognize that conventional solar equipment is, ordinarily, too expensive and that cheaper approaches must be found. Above all it is meant for people who wish to inquire beyond the conventional wisdom of solar heating—people who enjoy exploring in a strange territory filled with prickly, undomesticated ideas guaranteed by no one: good ideas that may be hard to recognize and bad ideas masquerading as good ones.

Persons wishing to learn about proven systems should avoid this book. Most of the well-known and thoroughly proven systems are not discussed here. Emphasis is on radically new approaches, which may or may not turn out to have real merit.

CENTRAL GOAL: CUTTING COST

Throughout most of this book emphasis is on cutting cost, i.e., inventing solar heating systems the overall costs of which will be very low. In some instances, emphasis is on performance: improving performance even if this entails some added complexity and some added cost.

WHO INVENTED THE SCHEMES DESCRIBED?

Most readers will care little who invented the schemes described. In a few instances the inventions have been made independently by several persons. Many of the inventions were made by me; but I presume that, in many instances, I have been preceded by other inventors.

In some instances, other persons made the inventions independently *after* I did. In a few of these cases they went further and patented the inventions. Just how valid such patents are, I do not know.

I have made no great effort to ascertain who were the first true inventors of the inventions described; but, ordinarily, whatever information I have at hand on such inventions I have included.

REASON FOR INCLUDING DESCRIPTIONS OF CERTAIN INVENTIONS I HAD NO HAND IN

It sometimes happens that the inventor of an excellent solar heating scheme fails to write a clear description of it. Few other persons learn about it and practically no one understands it. In a few instances I myself have prepared a description of the scheme, usually with the help and encouragement of the inventor.

Many of the schemes I have invented are merely improvements on other persons' schemes, and it has seemed appropriate to describe their schemes briefly before showing the improvements I propose.

ARE THE INVENTIONS PATENTED?

With one exception, I have not patented or applied for patents on any of the schemes described here.*

The one exception, discussed on page 202, is a brute-force method of overcoming the well-known obstacles to the storage of a large amount of heat in a large tank filled with Glauber's salt or various other phase-change materials. Broad Corporation of 73 Tremont St., Boston, MA, took the initiative in preparing a patent application on this invention; they paid the fees involved, and they own the patent (US Patent 4,117,882).

Whether or not other persons have applied for patents, or actually obtained patents, on the various inventions made by them or by me, I do not know, in most cases. I have made no systematic attempt to find out. I have, however, browsed through some listings of recent patents on solar energy, and I have looked over these two books:

Stanley Garil, *1977 Solar Energy Inventions and Design Patents*, 1978, paperback, 66 pp., $10. Available from PO Box 50003, F St. Station NW, Washington, DC 20004.

J. K. Paul *Solar Heating and Cooling 1977: Recent Advances*, 1977, cloth, $48. Available from Noyes Data Corp., Mill Road at Grand Ave., Park Ridge, NJ 07656.

SERIAL NUMBERS AND DATES OF INVENTIONS

Many of my inventions of collectors have serial numbers, such as C-20, C-30. Many inventions of entire solar heating systems have numbers such as S-20, S-30. The assignment of numbers was

---

*I made several inventions many years ago, and many of these were patented. Appendix 1 lists patents obtained in the period 1943—1966.

4

somewhat arbitrary, but some effort was made to give adjacent numbers to schemes of the same general kind.

Typically, the dates indicated are the dates when the invention was first written up carefully. In some instances, I have recently revised the material slightly to improve the clarity, to make the terminology more consistent, or for some other reason.

I have not assigned serial numbers or dates to inventions not made by me.

PUBLICITY ALREADY GIVEN TO MY INVENTIONS

Typically, after I have written a report on an invention, I mail copies to 10 to 50 friends, experts, and publishers. I do so in the hope that these persons will enjoy learning of the inventions, will make interesting comments on them, and will publish them. I have received many interesting comments. Some of the reports have been reproduced in Solar Energy Digest, various national or local solar energy newsletters, and other periodicals.

In summary, a few of my inventions have already been published formally and most have been published in an informal and limited way. A few are brand new.

FAVORITE STRATEGIES IN INVENTING SOLAR HEATING SYSTEMS

Strive for simplicity.

Design for tomorrow's austere living conditions, not today's luxurious conditions.

Use a costly component of a solar heating system for several purposes. (I am indebted to N. B. Saunders for recommending this strategy.) For example, use a massive wall not only as a wall but as a thermal storage element. Use a small storage system as a table, counter, or daybed also. A reflective panel that is used in summer to exclude solar radiation can be used on sunny winter days to reflect additional radiation into the rooms and can be used on cold nights as a thermal shutter.

Always use at least a moderate amount of passive solar heating.

Use a mix of passive and active solar heating.

Avoid mounting heavy and delicate equipment on hard-to-reach sloping roof.

Use simple manual controls rather than complicated, high-technology automatic controls. *Simple* automatic controls, of course, are entirely acceptable.

Use systems that are understandable and can be serviced by anyone handy with tools.

ATTEMPT TO MAKE THE DESCRIPTIONS STRAIGHTFORWARD AND SPECIFIC

I have tried to make each account follow a single straight line, i.e., tried to describe a single specific embodiment of the main idea. It would be all too easy to indicate multiple options, as in this sentence: "Connect the panels in series or parallel, transfer the heat—with or without a heat exchanger—to a water-filled tank of steel or fiberglass situated in the attic or basement or garage." Such a list of alternatives obscures the essence of the invention and bewilders the reader.

Patent lawyers often take the opposite approach. They try to expand a concise invention into a vast range, or field, of inventions. They employ vague words suggestive of broad classes of functions and wide ranges of uses. As a consequence, the reader may be unable to discover the essence of the invention. How is he to know that Smith, claiming "a rigid, arc-shaped, manually operable, control yoke" has merely invented a handle for a bucket?

After I have presented the central idea of an invention in terms of a specific embodiment, I sometimes list a few attractive modifications or embellishments. Usually, many more modifications could be proposed; but this would make the book too long.

UNCERTAIN PRACTICALITY OF THE INVENTIONS

Most of the inventions made by me have not been tried out. Would they really work? Would they work well? I cannot be sure.

Typically, many details as to materials and design have not been worked out. Accordingly no careful, quantitative estimates of performance can be made.

Also, I do not know how much the systems would cost. The costs would be low, I think. But I do not know *how* low. In some instances, where the emphasis has been on improving performance rather than cutting cost, costs may not be low; but *cost relative to performance* would be favorable, I believe.

I think the most useful question, ordinarily, is not, "Is the scheme likely to be truly successful?" but "In what special circumstances would the scheme be successful?" A certain system might be highly successful in a large office building in a city, and unsuccessful in other applications. Another system might be a great success in a small cottage in South Carolina but a great failure in a ten-room house in Maine.

Warning: In summary, readers are warned that most of the inventions described here are to be regarded as ideas or suggestions. Typically, there has been no detailed design, no cost estimate, no estimate of amount of energy collected and stored, and no prototype construction.

ARE THE INVENTIONS NEW AND RADICALLY DIFFERENT FROM EARLIER SCHEMES?

Most are. Some are not; they are included because of their special merit, their historic importance, or because they set the stage for some of the newer ideas discussed. My impression is that more than 50% of the inventions described are really new and differ considerably from earlier schemes.

COMMENTS AND SUGGESTIONS WELCOMED

Many readers will see at once how some of the proposed schemes could be improved—could be made cheaper, made to perform better, or applied to a different problem with greater success. I would welcome hearing from such readers. Also I would be glad to learn of actual construction and trial use of any of the schemes described. Finally, I would be glad to learn of other promising, little-known inventions in solar heating.

DO THE SYSTEMS CONFORM TO GOVERNMENT STANDARDS?

In many instances, the solar heating schemes described here may not be covered by, or may not conform to, state and or federal standards. The standards now in force or now being drafted are usually applicable mainly to conventional solar heating systems—yesterday's systems.

New systems may be very special—very hard for standards writers to cover.

Passive systems may be especially hard to cover.

Combination passive-and-active systems may be almost impossible to cover.

Most difficult of all to cover are systems that are intimately incorporated in, and integral with, the house proper, and systems the components of which perform multiple functions. How, for example, can standards writers deal with a large insulating panel that serves on a sunny winter morning as a reflector to direct additional radiation into the rooms, serves on a cold winter night as a thermal shutter, and serves in summer to block incident solar radiation? To express the performance, or output, of such a device in quantitative terms is almost impossible, and to write standards that embrace all of the functions clearly and judiciously may be an impossible task.

Fortunately, the penalty for not conforming to standards is small. It consists, usually, of foregoing certain tax benefits. In any event, such benefits may be small, and the advantages of installing a truly cost-effective solar heating system may far outweigh the advantage of a tax benefit.

Some strong criticisms of government standards on solar heating equipment are presented in Part 10.

Safety is a very different matter. I assume that all of the systems described here could be designed so as to be safe.

WHY HAVE OTHERS NOT WRITTEN SUCH A BOOK?

A typical inventor, having hit upon a promising idea, proceeds to build a working model of the device. This takes much time. Also, he may obtain the help of a patent attorney and, after a year or two, file a patent application. While preparing the patent application, he feels it prudent to say little about the invention.

Later, he may see how the invention can be improved. He may consider filing additional patent applications. At this stage, too, he may remain secretive.

Thus, in general, inventors, being busy and secretive, are not inclined to write books describing their inventions. At best, they are likely to postpone "telling all" until many years after the inventions were first conceived. Especially if the inventors are associated with industrial corporations, a prompt public disclosure of inventions is frowned on.

When an inventor, or the corporation he works for, does at long last decide to publicize an invention, the temptation to overstress the good features and gloss over the bad features is almost irresistible. If the task of preparing the description is assigned to a public relations department or sales department, the description is likely to be highly slanted.

The available books (listed on a previous page) that list patents on solar heating systems, quote portions of the patents, and present photoreduced copies of the drawings, are not of much help. The authors of such compilations have not studied the patents and learned their essences. Nor have they tried to impart such essences to the readers. (The task would be almost impossible: a given patent may have no essence—other than the claims, which consist of very long strings of very vague words. Also, a single patent may present 25 or 100 variations and alternatives.) For all these reasons such books are of limited interest except perhaps to patent attorneys.

OMISSION OF MATERIAL ON THERMAL SHUTTERS AND SHADES

I have included almost no material on thermal shutters and shades. These are of great importance. But they are considered in detail in my book *Thermal Shutters and Shades: Systematic Survey of Over 100 Schemes for Reducing Heat-Loss Through Large, Vertical, Double-Glazed, South Windows on Winter Nights*, draft edition, published Nov. 11, 1977, 19 Appleton St., Cambridge, MA 02138; $12.

WARNING CONCERNING ACCURACY

Material involving others' inventions, activities, patents, etc., may contain errors. No systematic attempt has been made to verify such material. No reliance should be placed on the information presented unless it has been verified independently.

WARNING CONCERNING PATENTS AND TRADEMARKS

Unless a reader has clear proof to the contrary, he should recognize the possibility that a given invention has been patented and a given name has been trade-marked. In most instances my information on these matters is inadequate and I have not attempted to make the situation clear to the reader.

Even when a writer states that his invention is patented, one may wonder whether the patent has really been issued or merely applied for, and one may wonder exactly which aspects of the invention are covered. Also there may be questions as to whether a court will eventually declare the invention to be trivial and the patent to be invalid. In some instances the patent may already have expired: its 17-year life may be at an end.

When a writer states that a name he employs has been trade-marked, one may wonder whether this is true, i.e., may wonder whether the application for trade-mark has been officially examined and approved.

WARNING CONCERNING ESTHETICS

Many of the solar heating systems proposed here entail changes in the appearance of the house. The vertical south face of the building, the south-sloping roof, or the south portion of the living room may have a strange new look. Conservatively minded people may dislike this. But when the looming energy crisis worsens, most people will, I think, regard anything that is nicely suited to keeping the house warm in winter as beautiful. When the going gets tough, beauty and functional excellence become almost synonymous.

ORGANIZATION OF MATERIAL

In general, I have tried to present the simplest and most interesting material first, and the complicated and dull material last. Thus, I start with non-concentrating systems: new kinds of passive systems, passive-and-active systems, and purely active systems. Then new kinds of systems that concentrate the solar radiation are described. Next come new kinds of storage systems: systems employing conventional materials (stones, water) and phase-change materials. Also included are some essays on terminology, test methods, popular fallacies, government standards, a comprehensive factor-of-merit for comparing solar heating systems. There are a few essays that are meant to be funny; DOE and HUD officials should avoid these.

ACKNOWLEDGMENTS

I am indebted to many persons for explaining their inventions to me, for criticizing my inventions, or for pointing out how various inventions could be improved. I am particularly indebted to Steven C. Baer, Maria Telkes, and John C. Gray.

PART 1

Passive Systems

INTRODUCTION

The number of passively solar heated buildings is increasing fast. My impression is that the cumulative number of such buildings in the USA is doubling every six months. Conferences on passive systems are oversubscribed. Lecture halls are packed. Several books and countless articles on the subject have been published recently.

Definition

If a solar heating system does not employ machines for circulating liquid through pipes or air through ducts, the system is said to be passive. Note that the system may, nevertheless, make use of shutters, dampers, vanes, or curtains that are operated at the start and finish of the day and may employ fans to help circulate air in an informal manner. (Some other writers employ different definitions.)

Status

Passive solar heating is an assured success. Typical systems are simple, reliable, understandable, and they supply much heat at low cost. Often they provide extra dividends in the form of greenhouse space or a welcome increase in room-air humidity in winter.

But there are also some troubles. The amount of heat that is stored may be too small; heat input and output may be too sluggish; the swing in room temperature (from 55°F to 80°F, for example) may be too large. On sunny days the south rooms may have excessive glare. Thus there is a crying need for improved designs of passive systems.

Is Passive Solar Heating Out-of-Bounds for Big Industry?

Scores of prestigious companies are busily producing components for solar heating systems of active type. They are producing collector panels, storage tanks, pumps, control systems, etc. But few, if any, are deeply involved in passive solar heating.

Why is this? Why are prestigious companies turning their backs on passive solar heating? I see these possible answers:

1. They may feel that passive systems are so crude, and allow such large swings of temperature, that discriminating people would not be willing to live in houses having such systems. Industry should not associate itself with such a crude way of life.

2. They may feel that passive solar heating systems are appropriate only for small houses situated far out in open country (no trees, no nearby buildings). Thus they may believe that the potential market for passive equipment is small.

3. They may feel that there is no way for industry to contribute to passive systems. They may believe that passive systems require only intelligent designs by architects and intelligent construction by local builders.

I believe that, while there is some truth in these views, there is a large role that industry can play in passive solar heating. I believe that many of the passive systems built in the next decade will include some high-technology components and that these will improve performance. Passive systems do not necessarily have to be crudely designed and executed; they do not necessarily have to allow large temperature swings.

Passive solar heating is in its infancy. There are many ways in which industry could help it prosper. Industry could help by developing better window structures. For example, it could develop windows with plastic glazing films that have lower reflection losses, with controlled air-inflow heat recovery, with thick reflective plates that can be used on sunny days to reflect additional amounts of direct radiation into the building, at night as thermal shutters, and in the summer as awnings to exclude solar radiation. Also, industry could develop walls and floors that have greater thermal capacity and faster heat input and output. Containers filled with water or with an appropriate salt hydrate could be incorporated in the walls and floors.

In the following pages, dealing with passive systems of non-concentrating type, I consider first the Trombe wall, which has been widely discussed and widely used. I list its shortcomings. Then I describe ways of improving it—minor and major improvements. Some of the proposed improvements have been tried out; but most have not. Then I discuss several other kinds of passive systems.

TROMBE WALL: ITS SHORTCOMINGS AND SOME POSSIBLE MINOR IMPROVEMENTS

THE SHORTCOMINGS

The Trombe wall, although successful in intercepting much solar radiation, storing a modest fraction of the energy received, and delivering a moderate amount of energy to the rooms after an appropriate time delay of 5 to 10 hours, has several shortcomings. The main shortcomings are:

Much of the energy that enters the wall soon leaves it—escapes southward from the south face. Much of this energy continues on through the glazing to the outdoors and is forever lost from the building.

There is no simple, effective, and inexpensive way to interpose, between wall and glazing, a thick insulating layer that will stop heat-loss on cold nights.

There is no simple way by which the house occupants can gain access to the space between wall and glazing for the purpose of cleaning and repairing the glazing.

On cold sunny mornings, when the rooms are too cold, little or no solar radiation can enter the rooms directly and, consequently, there is no quick warm-up.

There is no simple and effective way to stop the flow of heat from wall to the rooms when the rooms are already too hot.

The wall blocks off light from the rooms. Even on a bright sunny day the rooms may remain gloomy.

The wall greatly restricts the room-occupants' view of the outdoors.

In summary, the wall performs a few desirable functions fairly well but performs also several undesirable functions. It is a "good try," but not good enough.

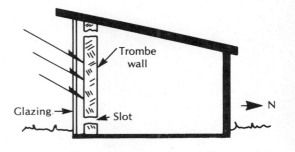

Vertical cross section, looking west

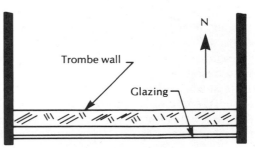

Horizontal cross section

MINOR IMPROVEMENT INVOLVING VERTICAL FINS

One way to somewhat improve the performance of the Trombe wall is to equip the south face with a set of vertical fins, perpendicular to the wall, black on the east sides and specularly reflective on the west. Such a scheme has been proposed by the Environmental Research Laboratory of the University of Arizona. See *Proceedings of the 2nd National Passive Solar Conference*, Vol. III, p. 910.

In the morning most of the direct radiation strikes the east (black) sides of the fins, warming them. Gravity-convective flow of room air picks up the heat from the fins and carries it into the room. Thus the room is warmed long before any substantial fraction of the wall has become warm.

At midday and in the afternoon, most of the direct radiation reaches the black wall proper—either directly or after reflection from the west (reflective) sides of the fins. Thus during this period the rooms, which are already warm enough, receive little heat and the wall becomes fairly hot and can keep the room warm at night.

Some drawbacks to this scheme are: (1) the added complexity and cost, (2) the amount of space taken up by the set of fins—perhaps necessitating placing the wall farther from the glazing, (3) interference between the set of fins and whatever shutter or shade is to be installed each night between wall and glazing. It is unfortunate that the set of fins cannot be made to serve also as nighttime shutter.

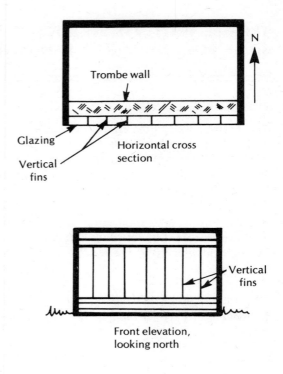

Horizontal cross section

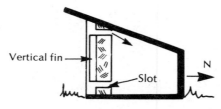

Front elevation, looking north

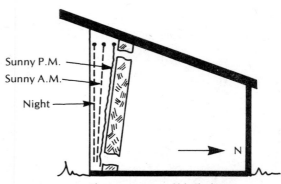

Vertical cross section, looking west

MINOR IMPROVEMENT INVOLVING A BLACK PLASTIC SHEET

By making appropriate use of a large black plastic sheet, one can modify the Trombe wall system so that in the morning most of the solar energy received is delivered promptly to the rooms and in the afternoon most of the energy remains in the wall—for use in warming the rooms some hours later.

The black plastic sheet, which is large enough to cover the entire south face of the Trombe wall, is installed between the wall and the window. The sheet is suspended from a long, stiff horizontal bar, which can be moved a few inches north or south by means of pulleys and ropes.

Three positions of black sheet between sloping Trombe wall and window

During the morning the sheet is suspended 2 inches from the wall; the sheet absorbs solar radiation, becomes warm, and heats air that circulates by gravity convection (via slots in the wall) to the rooms. During the afternoon the sheet is pressed tight against the wall so that most of the energy absorbed by the sheet flows directly into the wall. At night the sheet is suspended close to the window and thus helps define a thin region of trapped air that reduces heat-loss through the window.

How does one arrange to have the black sheet press firmly against the Trombe wall? If the wall slopes upward to the north (rather than being vertical), gravity will press the sheet against the wall. A more positive method is to provide a network of grooves in the south face of the wall and apply (by means of a very-low-power blower) steady suction in the space between sheet and wall; thus the sheet is everywhere pressed against the wall by atmospheric pressure from the south.

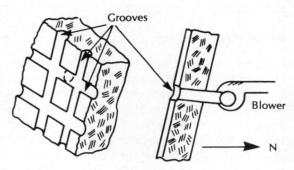

Portion of sloping Trombe wall that has been grooved so that a black sheet of plastic can be held close against it by suction.

COMMENT

It seems to me that neither of these "thought–experiment" improvements of the Trombe wall is outstandingly successful. Each is helpful, but each adds to the complexity and cost. Each deals with only a few of the shortcomings listed above.

14

SCHEME INVOLVING DIVIDING THE TROMBE WALL INTO SEVERAL SMALL WALLS EACH TURNED 10 DEGREES

Scheme S-179
12/16/77

PROPOSED SCHEME

Here we propose a major redesign of the Trombe wall: a major change that obviates nearly all of the usual shortcomings. The wall is divided into pieces about 5 ft. long. Each piece is rotated so as to make an angle of about 10° with the window (in plan view). Each piece is tapered, being 6 in. thick at the top and 12 in. thick at the bottom. For each piece, double tracks (attached to ceiling) are provided on the south side of the piece and another set of double tracks is provided on the north side. Each track runs approximately in a straight line along the south side of one piece and along the north side of an adjacent piece. Heavy, insulating, ceiling-to-floor curtains (each about 6 ft. in E–W dimension) are suspended from each track.

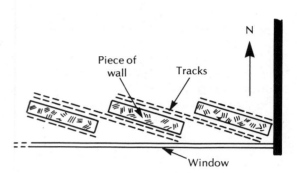

Plan view showing three pieces of wall and several pairs of tracks for curtains

OPERATION OF THE SYSTEM

On a cold night all of the curtains are slid along so as to be situated between wall-pieces and window, that is, there is a double layer of insulation on the south face of each piece and no insulation on the north face; thus little heat is lost to the outdoors and much heat can flow into the room. Early in the morning of a sunny day in winter the curtains are slid along so as to be situated on the *north* sides of the pieces, and 1-ft.-wide, ceiling-to-floor gaps remain between pieces; thus, a moderate amount of solar radiation can enter the room, via these gaps, to warm it promptly, there is good daylight illumination, and the occupants can enjoy a view of the outdoors. At midday on a sunny day in winter the curtains remain on the north sides of the pieces, so that the pieces become hotter and hotter and store more and more energy. At dusk on a winter day the curtains are distributed on both sides of the pieces; thus loss of heat to the outdoors is kept fairly low and flow of heat to the (already hot enough) rooms is kept fairly low. At 9:00 p.m. on such days the curtains are adjusted so that all are between wall-pieces and the window.

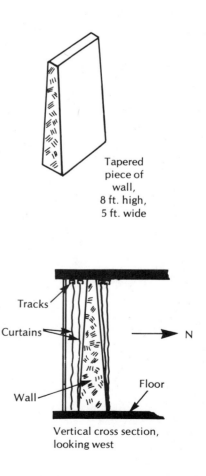

Tapered piece of wall, 8 ft. high, 5 ft. wide

DISCUSSION

The reason for tapering the wall-pieces, i.e., making them thinner at the top than at the bottom, is that the curtains will then be pulled by gravity tight against the faces of the pieces. The tracks are situated close in, and accordingly the curtains press against the pieces. Thus effective insulation is achieved irrespective of the fact that there are no edge seals. Almost any amount of pressure can be achieved if the lower part of the curtain is suitably weighted, e.g., by a horizontal steel chain affixed 3 in. above the floor.

Vertical cross section, looking west

Notice that because of the rotation of the wall-pieces occupants can reach into the space between the wall pieces and the window, for the purpose of cleaning the window, for example.

On very cold nights it will pay the occupants to close the gaps between wall-pieces so that practically no cold air from the space close to the windows can flow into the room. (That space will become especially cold, because little heat from the wall flows into it, thanks to the two layers of heavy curtains.) To close the gaps one may use closure devices that join the pertinent curtains edge-to-edge, or one may employ insulating flaps that are attached to the ends of the pieces, e.g., by means of hinges or sliders.

The proposed scheme would work well in summer also, especially if the south faces of the more southerly curtains were aluminized or white in color.

REPLACING THE TROMBE WALL WITH A ROW OF 4-FT.-HIGH, RECTANGULAR, WATER-FILLED TANKS EQUIPPED WITH THREE-MODE INSULATING PLATES

PROPOSED SCHEME

Here we propose replacing the usual Trombe wall with an east–west row of rectangular, obliquely mounted, black-painted, water-filled steel tanks. Each tank is 6 ft. long, 1 ft. thick, and 4 ft. high. Inasmuch as the tank tops are at chest height, the occupants of the room can look out over the tops of the tanks to enjoy the view. Also, some direct solar radiation passes above the tanks and penetrates deep into the room, helping warm it. Throughout the day much diffuse radiation enters via the space above the tanks, illuminating the room.

Each tank is mounted slightly obliquely so that the north face of one tank is nearly coplanar with the south face of the next. Thus a large insulating plate that, by day, is parked along the north face of one tank may, at 6:00 p.m., be slid along so as to insulate the *south* face of the next tank. The plates slide in channels, which insure accurate positioning of the plates and also provide edge seals.

In all, there are *two* insulating plates that can lie close to the north side of a given tank. One can be slid so as to serve the neighboring tank, as indicated above, and the other used to control the northward flow of heat from the given tank; at night when the room threatens to become too cold, this plate is slid out of the way so as to allow the north face of the given tank to deliver heat to the room by radiation and convection. (This plate is slid out of the way by sliding it in the same direction and same manner as the first-mentioned plate.)

The tank ends and tank tops are permanently insulated. Books, flower pots, etc., can be placed on the tank-top insulation; that is, the tank-top insulation can be used as a shelf or counter.

Notice that the set of tanks collects much energy, stores much energy, and is equipped with a versatile and effective insulating system. Also, the room receives some direct radiation and receives good illumination all day.

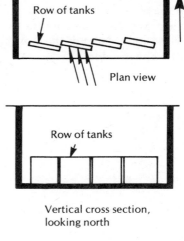

Row of tanks

Plan view

Row of tanks

Vertical cross section, looking north

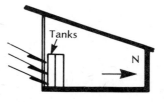

Tanks

Vertical cross section, looking west

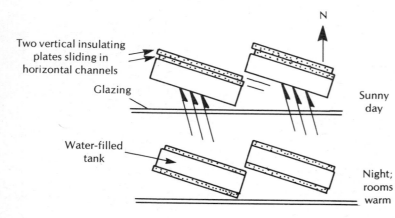

Two vertical insulating plates sliding in horizontal channels

Glazing

Water-filled tank

Sunny day

Night; rooms warm

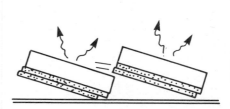

Night; rooms need heat

Plan-view of two tanks

BIER'S SET OF FIVE SMALL ABSORBER-AND-STORAGE WALLS AT 45 DEGREES

While I was inventing Scheme S-179, described on previous pages, James Bier of Ferrum, Virginia, was already building a small house incorporating a somewhat similar scheme. His invention preceded mine. A detailed description of his house is presented in my book *Solar Heated Buildings of North America: 120 Outstanding Examples*.

Bier's two-story house, with massive walls and floors, makes much use of passive solar heating. Nearly the entire south face of the house is (in winter) double glazed; immediately north of the glazing there is a row of five concrete walls. Each wall is 15 ft. high and serves both stories. In horizontal cross section each wall is 32 in. by 12 in. Each is at 45° to the south wall; that is, it lies in a southeast-northwest plane. The five walls, 4 ft. apart on centers, serve as a set of optical louvers and as a storage system. Specifically, they (1) admit much light to the rooms, (2) permit occupants to obtain a view to the southeast, south, and southwest (but not to west-southwest), (3) allow much morning solar radiation to penetrate deep into the rooms, warming them, (4) intercept and absorb most of the afternoon radiation, thus reducing the tendency for the rooms to become too hot, (5) store much heat in the afternoon, (6) release much heat at night, thus keeping the rooms warm, (7) serve as supports for vertical curtains or shutters that, at the end of the day, may be slid into position to insulate the large south windows, (8) help support the roof. Note that the set of five walls serves the main function of a Trombe wall but avoids most of its drawbacks.

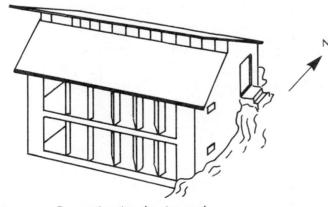

Perspective view showing south window-wall, set of five short masonry walls, and clerestory windows

Plan view showing the 45° orientations of the five short masonry walls

SIMPLE, CONTROLLABLE INSULATION FOR THE TWO FACES OF A 7-FT.-HIGH MASONRY WALL AT SOUTH SIDE OF SOLAR HOUSE

PROPOSED SCHEME

A 7-ft.-high masonry wall situated indoors close to the big vertical south window of a passively solar heated house is controllably thermally insulated by means of a large thick blanket that can be slid up and over the top of the wall, so as to insulate the north side of the wall during the daytime and (after it is shifted) the south side at night.

Wall

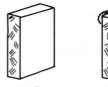

Wall and blanket

If the top of the wall is rounded and smooth, the blanket may slide up-and-over easily enough. Of course, pull-cords are attached to each end of the blanket so that the room-occupant doesn't have to climb a ladder to pull down the upper end of the blanket.

If the top of the wall is not rounded and smooth, one installs slender rollers along each of the top edges of the wall. The rollers are mounted on simple bearings or on ball-bearings.

To make sure that the blanket cannot go *completely* up and over (with the result that one cannot reverse the process), one makes the pull-cords amply long. Also, one may attach 10-lb. weights to the ends of the cords to insure that these ends will never leave floor-level. Or one can simply fasten the cord-ends to the floor.

Blanket on right

Blanket on left

If the wall is tapered slightly, being wider at the bottom than at the top, gravitational forces acting on the blanket will tend to keep it pressed close to the wall. If it were not close and there were no edge seals, the effectiveness of the insulation would be greatly reduced.

Suppose one would like *neither* face of the wall to be insulated. How can this be arranged? Several schemes are available:

1. Fold the blanket once or twice to reduce its effective area, then hoist this folded assembly so that it resides on, or close to, the top of the wall, leaving major fractions of both faces exposed.

2. Allow the blanket to strike the floor and fold upon itself there, leaving both faces of the wall exposed.

3. Arrange the roller systems so that each roller can be moved 6 in. away from the wall, i.e., so that the blanket section suspended from it is kept well clear of the wall, allowing air to circulate freely between blanket and wall.

Tapered wall with rollers

Pullcord

10-lb. weight

Pullcord and weight

THREE-MODE REFLECTING-AND-INSULATING PLATES FOR USE WITH BIER'S ABSORBER-AND-STORAGE WALLS

Scheme S-181
8/4/78

PROPOSED SCHEME

I propose to improve Bier's system so that it will receive and store more energy and so that the flow of heat from walls to the room can be controlled.

In the proposed scheme each of the small storage walls is oriented at 60°, instead of 45°, from the big south window. A hinged, reflecting-and-insulating plate is installed on the NE side of each such wall and a detachable insulating plate is installed on the NW end of the wall.

At night the hinged plates are swung so that they close against, and insulate, large portions of the big south window. The detachable plates are mounted so as to fill the gaps, i.e., complete the insulation of this window. All of the faces of each storage wall are now bare: free to distribute heat to the room.

In the morning each hinged plate is swung back against the NE face of its associated wall, insulating it and allowing much clear space for the morning solar radiation to pass through and warm the room promptly. Also, the detachable plate is attached to the NW edge of the wall.

In the afternoon, starting at about noon, each hinged plate is swung through about 80° or 90° so as to form a reflector ideally oriented to direct much solar radiation toward the neighboring wall. Thus that wall receives much afternoon radiation directly and much indirectly, and almost no radiation penetrates deep into the room, which by now is likely to be hot enough.

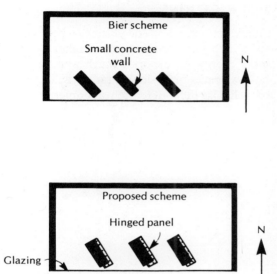

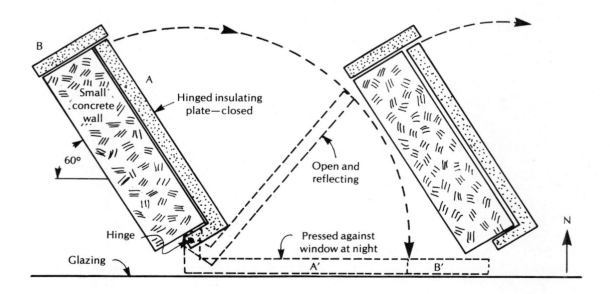

DISCUSSION

Much energy is collected and stored. It can be stored for relatively long periods. The rooms will very seldom become too hot. Night-time heat-loss through the window is greatly reduced. The big hinged insulating plates do triple duty.

MODIFICATIONS

Apply permanent plastic glazing to the SW face of each little wall, so that the wall may reach a higher temperature and thus store more energy. Incorporate much water in the wall (in plastic-lined hollow spaces or in slender tanks) to increase the thermal capacity of wall. Use the wall to preheat domestic hot water.

SET OF ABSORBER-AND-STORAGE WALLS EACH OF WHICH CONSISTS OF A SIDE-BY-SIDE PAIR OF VERTICAL, WATER-FILLED CYLINDERS WITH TWO-MODE, REFLECTING-AND-INSULATING PLATES

Scheme S-180
7/31/78

SUMMARY

Here I propose a further improvement on the Bier system: I propose to further increase the amount of heat stored, and I propose to increase the ease of heat input to, and heat output from, the special walls, or louvers.

The concrete walls are replaced by a spaced set of vertical louvers each of which consists of a pair of tall, 1-ft.-dia., water-filled tanks and a versatile housing that encloses the pair of tanks. In the afternoon, reflective doors on the SW sides of the housings are open and allow direct and reflected solar radiation to reach the tanks and be absorbed by them. In the evening, the housings confine the heat in the tanks until the room begins to become cold.

I would expect this system to perform twice as well as Bier's. But it is more complicated and expensive than his and requires much more attention from the occupants. Also, I would expect the proposed system to outperform nearly all conventional passive-solar-heating systems.

PROPOSED SCHEME

I propose that we modify James Bier's scheme by substituting a side-by-side pair of vertical, water-filled cylindrical tanks for each concrete louver and by providing multi-function housings for such pairs of tanks. Each tank is 12 in. in diameter, 8 ft. high, and is made of Kalwall Sun-Lite. It contains 380 lb. of water. Its exterior is black. (Such tanks are used in the Goosebrook House in Harrisville, N.H.) The plane that contains the centerlines of the two tanks makes an angle of 60° with a vertical E-W plane. (The Bier louvers are at 45°.)

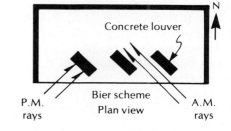

Concrete louver

P.M. rays

Bier scheme
Plan view

A.M. rays

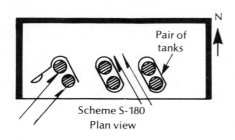

Pair of tanks

Scheme S-180
Plan view

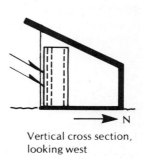

Vertical cross section, looking west

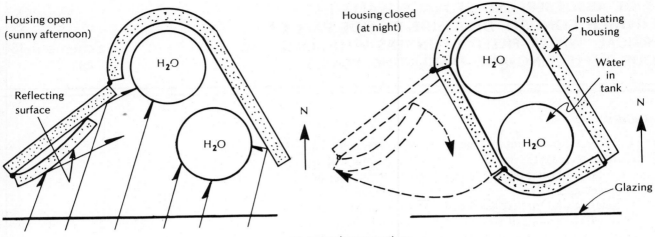

Housing open
(sunny afternoon)

Reflecting
surface

H₂O

H₂O

N

Housing closed
(at night)

Insulating
housing

Water
in
tank

H₂O

H₂O

N

Glazing

Horizontal cross sections

Each pair of tanks is enclosed in a housing. The housing portion on the N and NE sides of the pair of tanks is of 1-in.-thick insulation and is fixed. At the bottom there is a large opening by which room air can enter the housing whenever the housing cap is removed to allow hot air to escape, at the top, into the room.

The SW face of the housing consists of a vertical, hinged, insulating, folding door. When the door is fully closed, it completes the insulation of the pair of tanks. But when it is open and swung wide, it allows direct afternoon solar radiation to strike the pair of tanks, and (because the pertinent faces of the door components are covered with reflective aluminum) it also reflects an additional quantity of solar radiation toward the tanks.

The housing embraces the pair of tanks very loosely. That is, air within the housing is free to circulate, by gravity convection, on all sides of each tank.

There are many pairs of tanks. All are close to the big south window. The pairs are 4 ft. apart on centers.

Many 8 ft. × 2 ft. × 2 in. insulating plates are provided and are used on cold nights to insulate the big south window. Because insulation is used, it is permissible that this window be merely single glazed; this cuts costs and increases the transmittance by about 10%. For each pair of tanks a pair of such plates is provided. They are clipped, side by side, to the window by one of the means described in my book *Thermal Shutters and Shades*. In the morning these plates are removed; they are hung (parked) against the fixed portions (NE faces) of the housings.

OPERATION

Until 11:00 a.m. on a sunny day, most of the solar radiation that passes through the big vertical south window passes between the housings and penetrates deep into the room and is absorbed there and helps warm the room promptly. At about 11:00 a.m. (or much earlier, if the occupant desires), the housing doors are swung open so as to (a) allow much direct radiation to strike the tanks and (b) reflect an additional amount of radiation toward them. A few of the doors may be kept closed or opened only half-way, if the occupant wants to have much light enter the room or wants a view. At midday and throughout most of the afternoon the tanks receive much solar radiation—directly and via the reflective doors. By 5:00 p.m. the temperature of the tanks may rise by about 12 F degrees (rise, say, from 75°F to 87°F).

At the end of the day, the occupant closes the folding doors, thus fairly well confining the heat within the housings. Also, he installs the insulating panels on the big south window.

When the room threatens to become too cold, he opens the top of the housing, thus allowing the hot air within the housing to rise and enter the room. Colder air, from near the floor, enters the housing via the always-open port near the bottom thereof. Additional circulation may be provided by partially opening the hinged door.

In milder months, most of the above-mentioned operations can be omitted.

DISCUSSION

Because the pair of tanks (louver) contains 760 lb. of water, it stores much heat. Heat input and heat output are very efficient, inasmuch as the water within each tank circulates freely by gravity convection. The mass of water is only about 1/3 the mass of the Bier concrete louver, but the specific heat of water is 5 or 6 times that of concrete. Thus the long-term thermal capacity of the pair of tanks is about twice that of the concrete louver and the effective short-term capacity may be 2½ to 4 times that of the concrete louver. Because the pair of tanks takes up more heat on a sunny afternoon than a concrete louver, the tendency of the room to become too hot in the afternoon is reduced. Because the pair of tanks may be left fully insulated throughout the first part of the evening, room overheating at that time is reduced, too. Because most of the heat in the pair of tanks can be confined there for many hours, thanks to the insulating housing, the amount available for use when the rooms become cold, i.e., 5, 10, or 20 hours later, is greater. And because the water inside the tanks is self-circulating, the amount of heat that is readily released by the tanks is especially great.

I would guess that the proposed system would work, overall, twice as well as Bier's. However, the proposed system is more complicated, more expensive, and requires frequent attention by the occupants if the best possible performance is to be achieved. Also, the louvers (each including two tanks and a housing) are 50% thicker than Bier's louvers; hence they take up more room and block the window a little more.

Bier's solar heating system is so very inexpensive that, it seems to me, adding a moderate amount to the cost is permissible. In general, passive systems (relative to active systems) are "more than cheap enough.'' The weak point is performance—rooms get too hot in the afternoon; too little heat is stored; there is no control over the stored heat. To go to moderate expense to overcome these limitations may be well worthwhile.

MODIFICATIONS

Scheme S-180a

As above, except provide, at the housing inlet port, a very-low-power blower to assist the extraction of heat from the pair of tanks.

Scheme S-180b

Instead of including just two tanks within each enclosure, include a row of three tanks. Then the system will absorb and store about 50% more energy with negligible increase in complexity.

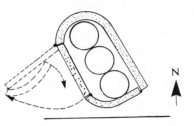

Scheme S-180c

Modify one or two of the tanks in such a way as to increase the stratification and increase the temperatures of the upper portions of the tanks; then tie these portions into the domestic hot water system. The system will then provide a fair fraction of the energy needed for domestic hot water, as well as providing a large fraction of the energy needed for space heating.

Scheme S-180d

As above, except, additionally, make the insulating panel on the NE face of the louver openable to facilitate the release of heat to the room. (I am indebted to J. C. Gray for suggesting this improvement.)

HUNT'S ROW OF VERTICAL, CYLINDRICAL, WATER-FILLED ABSORBER-AND-STORAGE TANKS

HUNT'S SYSTEM

The system, used in the house of M. B. Hunt of Davis, Calif., bears a resemblance to the Trombe wall but seems superior to it in many respects. In Hunt's house, half of the vertical south wall consists of a large (14 ft. × 14 ft.) double-glazed window area. Immediately north of this area there is a row of seven tall, slender, black water-filled tanks. Each tank, of spirally-formed galvanized steel, is 18 in. in diameter and 14 ft. tall. Each contains 24 ft^3 (1500 lb.) of water. Total quantity of water: 5 tons. Solar radiation strikes the tanks directly, warming them. They store much energy. The 6-inch gaps between the tanks permit natural illumination of the room and provide some view of the outdoors.

DISCUSSION

The system is said to perform very well in this location where the climate is moderate.

I suppose that it would work well in colder climates also, provided that simple housings were provided for the tanks in order that the flow of heat from tanks to rooms could be controlled. Perhaps thermal shutters or shades should be provided also.

A detailed description of the house is presented in my book *Solar Heated Buildings of North America: 120 Outstanding Examples*.

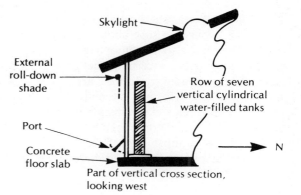

Part of vertical cross section, looking west

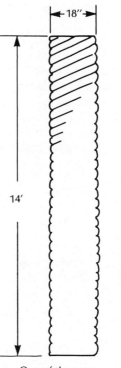

One of the seven vertical tanks

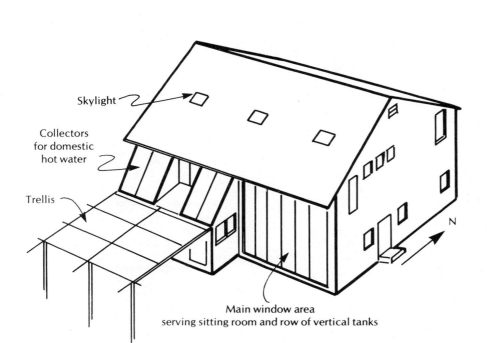

OUTDOOR NEAR-HORIZONTAL REFLECTOR
MOUNTED AT BASE OF SOUTH WINDOW

SUMMARY

A surprisingly large amount of supplementary solar heating of the south rooms of an existing house in a sunny location can be provided by installing, just outside the sills of the south windows, crude, flat, slightly-sloping, plywood-and-aluminum-foil reflectors. Using several such reflectors, costing about $200 in all (in 1973), the owner can save about $50 to $100 of purchased energy each winter and also can enjoy extra warmth in those rooms and reduce his worries concerning fuel shortages and electrical blackouts. The proposed reflectors are cheap, durable, and easily installed. Their effective thermal capacity is zero: delivery of energy to the rooms starts instantly when the sunlight strikes the reflectors. The reflectors make *no* contribution to heat losses at night, etc. In summer the reflectors can be re-mounted so as to serve as blinds or awnings to exclude radiation and keep the rooms cool.

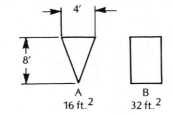

PROPOSED SCHEME

Figure 1 shows a conventional house with four types of reflectors mounted adjacent to south windows, here assumed 4 ft. wide and 5 ft. high. Figure 2 shows the (5°) tilt of a reflector.

Tilt of Reflector

For houses at latitudes such as that of Boston (42°N), the altitude of the sun at noon on a typical day in midwinter is, say, 30°, and at noon on such day a Type–B reflector (4 ft.× 8 ft.) that is tilted 1° or 2° downward-toward-the-south nicely fills the window with reflected sunlight. But an elevation angle of 25° is more nearly typical of a sunny hour on such day, and for this elevation angle a reflector tilt of 5° is about optimum.

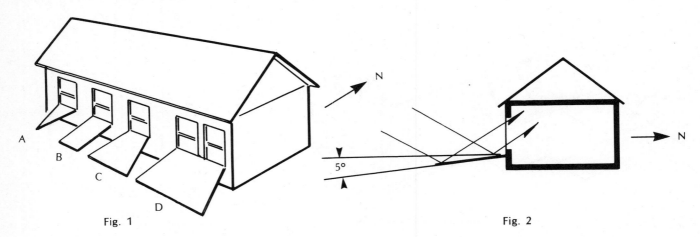

Fig. 1

Fig. 2

The Reflecting Surface

The reflector might consist of a ½-in. sheet of weatherproof plywood to which sheets of shiny aluminum foil or aluminized mylar have been affixed, by glue, tape, or other means. The owner could install a fresh reflecting sheet every few years, if necessary. Or he could use plywood to which a thin aluminum sheet had been bonded at the factory. Some benefit would be achieved using merely aluminum paint or white paint.

Relative Merits of the Four Types of Reflector

In terms of the amount of energy delivered per unit area of reflector, the Type-A device is better than B or C. All parts of it are so close to the window and so close to the north-south centerline of the reflector that they deliver energy to the window throughout a considerable portion of the day. Type B, although delivering more energy, is slightly less effective inasmuch as the outer corner parts contribute throughout a slightly shorter portion of the day. Type C, although delivering even more energy, has even lower effectiveness, because some parts are even farther from the window and from the centerline and, hence, are effective for brief portions of the day only (when the sun is so low in the sky that the effective projected area of the reflector is relatively small).

Where there are two windows close together, Type D can be used especially effectively. Nearly every part of the reflector delivers energy to one window in one portion of the day and to the other window in another portion of the day.

Direct Energy vs. Reflected Energy

It is worth noting that the amount of energy entering the windows directly from sun and sky is greater than the amount entering via the reflector, because of the reflector's imperfect reflectance, imperfect tilt, and limited area.

Curvature of Reflector

It would be slightly beneficial to bend the reflector slightly (make it slightly concave upward) and to allow the south corner portions to droop slightly in order to compensate for the low elevation of the sun at times (remote from noon) when these portions come into play. But such changes would probably not be worth the effort.

Amount of Energy Delivered by the Reflector

Some crude calculations I have made suggest that the four types of reflectors, A, B, C, and D, would deliver roughly 4000, 8000, 14,000 and 31,000 Btu per typical sunny day in the winter when applied to the above-specified south window of a house near Boston. These amounts correspond to about 1, 2½, 4, and 10 kWh.

Value of Energy Delivered
If the energy delivered by the Type D reflector were provided, instead, by electric heaters consuming electric power costing 4¢/kWh, the cost of such power would be (10 kWh) ($0.04/kWh) = $0.40 per day. If the reflector were used throughout the equivalent of 120 sunny days per winter, the energy supplied would be worth (120 day) ($0.40/day) = $48.

I estimate, similarly, that for A, B, C, and D the amounts of energy provided per winter would be worth (in terms of electric power cost) $6, $12, $20, and $48, respectively. Relative to energy obtained from burning oil, at the 1979 price of oil, the amounts of energy in question would be worth about half as much.

Cost of Reflector
The type-B reflector, consisting of a 4 ft. × 8 ft. panel of weather-proof plywood to which an aluminum coating has been affixed, would have a total cost of $25, I estimate. Costs of Reflectors A, B, C, and D might be about $15, $25, $45, and $65.

Money Saving
Reflector D would pay for itself in about 1½ years, relative to electric power, and in about 3 years relative to use of oil. It would take the other types of reflector about one half again as long to pay for themselves.

Other Benefits
The reflectors would give the residents satisfaction from (1) having one room that is about 10 F degrees hotter than typical rooms (i.e., 70°F instead of 60°F, say) on sunny days in the winter and (2) being somewhat less dependent on the continuity of oil supply and continuity of electric supply (essential for operation of an oil furnace). If several reflectors were used, the residents would have the satisfaction of knowing that, even if, in midwinter, the fuel supply or electric supply failed during a period in which there was a typical amount of sunny weather, the house would keep above 32°F and, accordingly, no water pipes would freeze and burst.

Note: On a sunny day in winter, seven Type-C devices (or three Type-D devices) would supply about 100,000 Btu, which is about 1/3 of the amount of energy the furnace would normally be called on to supply on such a day for a typical, well-insulated, two-bedroom house near Boston.

Desirability of Installing Additional Windows
A homeowner might find it desirable to install an additional (double-glazed) window beside "lone" windows or between a pair of windows that are 5 ft. apart to obtain more radiation directly through such windows and to permit installing Type-D devices to contribute further to the partial solar heating of the house.

Installing the Reflector
The reflector-edge closest to the house could be screwed, nailed, or tied to the window sill or to wooden blocks previous screwed to the

side of the house. The outer edge of the reflector could be fastened to stakes or posts, or it could be rested on (and tied down to) a fence, wall, or heavy horizontal log. The reflector should be sufficiently secure to be unaffected by winds.

Summer Use of Reflector
Instead of storing the reflector in a garage in summer, the homeowner might raise up the outer end so as to form a blind for the window. Or he could raise both ends so that the device would form an awning or sunshade. Thus, the room could be kept relatively cool even on a sunny hot day in summer.

Adjusting Tilt of Reflector
The residents might find it worthwhile to adjust the tilt of the reflector slightly, from month to month, to increase its effectiveness. In December, for example, greater tilt is desirable.

If the "south" side of the house in fact faces 20° west of south, the residents might find it worthwhile to install the reflector so that it has a slight (about 10°) tilt upward-toward-the-west.

Installing Reflectors on Second Story Windows
This too can work out well. Here, the reflectors will not interfere with shrubs or walkways, and they are less prone to be shaded by nearby trees. The reflectors may be made shorter to avoid unduly shading windows below. The outer end of a reflector can be supported by poles, long brackets, or tie-rods running from the overhead eaves.

Actual Sizes of Windows
Most windows of most houses are much smaller than the size (4 ft. × 5 ft.) assumed in the present calculations. Thus, smaller reflectors would be used and, preferably, more of them. Being smaller, they would be easier to handle and install, but the overall cost per unit area would be somewhat greater.

DISCUSSION

The system is cheap, simple, rugged, has no moving parts, requires no adjustments, and uses no electricity. It has zero effective thermal capacity—it starts delivering energy to the room the instant sunlight strikes the reflector. Even on somewhat overcast days it delivers some energy to the room. The system makes *no* contribution to heat losses.

But it does somewhat disfigure the house-exterior and the grounds adjacent to the south side of the house.

The owner may have to remove snow from the reflector several times a winter. Snow itself has very high reflectance, but because the reflectance is almost entirely of diffuse character, instead of specular, the contribution to room heating is less than half that of the proposed aluminum reflector.

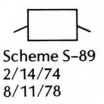

OUTDOOR VERTICAL REFLECTOR MOUNTED ADJACENT TO NORTH EDGE OF EAST OR WEST WINDOW

Scheme S–89
2/14/74
8/11/78

PROPOSED SCHEME

At several windows at the east end of the house and several windows at the west end of the house, install reflectors each of which consists of a flat, aluminized sheet of plywood. Each sheet is vertical and makes an angle of 30° with the vertical east–west plane. The width of the sheet is the same as the width of the window; the height of the sheet is the same as the height of the window. However, the midpoint of the sheet is 1 ft. higher than the midpoint of the window to allow for the fact that the sun's rays have a downward component. The reflectors are firmly attached; high winds will not affect them.

The reflectors at the east end of the house work fairly well from 9:30 a.m. until 12:15 p.m., and the reflectors at the west end work well from 11:45 a.m. until 2:30 p.m. The amount of energy delivered to the rooms is closely comparable to the amount delivered by a comparable area of aluminized plywood mounted horizontally near the base of a vertical south window. Note that the reflectors are cheap, light, and rugged, and start delivering heat to the rooms the instant the sun comes out from behind a cloud.

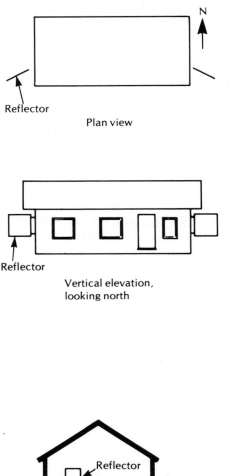

Plan view

Reflector

Vertical elevation,
looking north

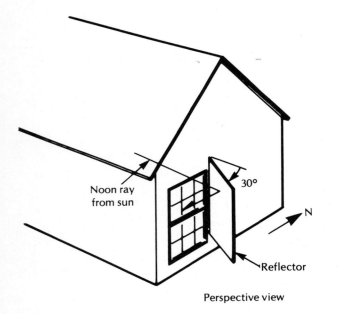

Perspective view

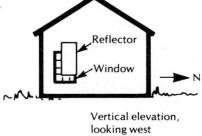

Vertical elevation,
looking west

MODIFICATIONS

Make the reflecting sheet slightly concave. Make it higher and wider. Make the height adjustable (very high in April, very low in December). Cut a horizontal slot in the sheet near the center, so that the occupants of the room can get at least a small view to the NE or NW. On cold nights swing the sheet so that it presses against the window, acting as a thermal shutter. In summer reverse the sheet: move the hinges to the south (instead of north) edge of the window, so that the sheet will exclude direct solar radiation. (The foregoing idea was proposed 9/8/78 by W. K. Langdon.) One could relocate the sheet so that it serves as an awning. Or, in summer, one could store the sheet in the basement.

IS A 45° ORIENTATION AS GOOD AS A STRAIGHT SOUTH ORIENTATION FOR A PASSIVELY SOLAR HEATED BUILDING IF USE OF EXTERNAL VERTICAL REFLECTORS IS PERMITTED?

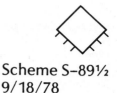

Scheme S-89½
9/18/78

SUMMARY

In many ways, a rectangular, passively solar-heated building oriented at 45° from the east-west direction can (I guess) outperform an equivalent building that faces straight south. The key to the success of the 45° orientation is the employing of many vertical reflectors.

INTRODUCTION

Everyone seems to have assumed that, ideally, a rectangular building that is to be passively solar heated should face south, as indicated in Figure 1. For a building that is not equipped with vertical reflectors, this assumption is presumably valid. But if vertical reflectors are used judiciously, the assumption may be wrong. A 45° orientation may be better, as explained below.

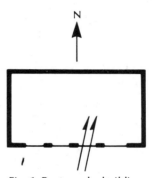

Fig. 1. Rectangular building that faces exactly south.

PROPOSED SCHEME

Figure 1 shows a rectangular, passively solar-heated building that faces exactly south, and Figure 2 shows a similar building that faces 45° from south. Each of the latter building's windows that faces SE or SW is equipped, along its north edge, with a vertical reflector mounted perpendicular to the window. Each such window receives much solar radiation for about 4½ hours every sunny day in winter; some radiation strikes the window directly, while an almost equal quantity strikes it after being reflected. In order that each reflector will not block radiation traveling toward the succeeding window, the windows should be of moderate width only and should be well spaced with opaque wall areas intervening. Inasmuch as the wall areas facing SE or SW considerably exceed the area of the south wall of Figure 1, one may estimate that the total amount of solar radiation entering the 45° building is roughly equal to that entering the south windows of the building that faces straight south.

Some interesting advantages of the proposed scheme are:

Solar heating is applied to two, not just one, sides of the building. Thus the distribution of solar heat is better. There is no region in which there are too many windows or too much glare, with the threat of local overheating.

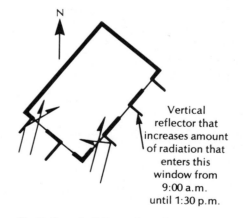

Vertical reflector that increases amount of radiation that enters this window from 9:00 a.m. until 1:30 p.m.

Fig. 2. Same building, oriented at 45° from south. Many windows are equipped with vertical reflectors.

A larger area of walls and floors receives the solar radiation; thus the effective thermal capacity is greater, carrythrough is greater, and overheating at 2:00 p.m. on sunny days is reduced.

In summer the reflectors can be swung so as to exclude radiation (but still admit enough diffuse radiation for illumination); thus, the building is kept cooler.

MODIFICATIONS

Scheme S–89½a

Increase the angular width of the south corner of the building: increase it from 90° to about 120°. This change considerably increases the amount of solar radiation collected. More specifically, it considerably increases the length of the period each day in which a given window collects much radiant energy. (I am indebted to J. C. Gray for pointing this out.)

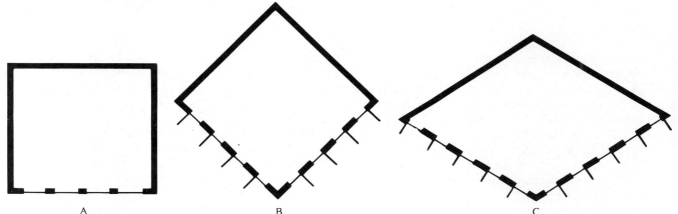

A   B   C

Fig. 3. Buildings that have the same floor area but have different orientations and/or different shapes. It is proposed here that, from the standpoint of passive solar heating, B slightly outranks A and C slightly outranks B. B and C employ vertical reflectors at all SE and SW windows.

The reflectors would perform better if they were extra–high, extra–wide and somewhat curved, or cupped. But the expense would be greater—too great, perhaps.

Scheme 89½b

Modify the reflectors so that they also serve to support an inflated, double–walled, transparent covering such as is sold by Solar Room Co. That is, apply inexpensive two–layer plastic covers to the SE and SW faces of the building, using the reflectors in place of the ribs, or in place of some of the ribs. The added expense might be repaid by the reduction in heat–loss through the pertinent building faces and other benefits provided by such solar enclosures.

Enclosure consisting of two sheets of plastic

Reflector that serves also as a rib to support the plastic sheets

Fig. 4. Building (at 45°) employing set of vertical reflectors within a transparent enclosure of plastic

TRANSPARENT, WALK-IN ENCLOSURE INSTALLED ALONG SOUTH SIDE OF EXISTING HOUSE AND USED TO HEAT THE HOUSE

INTRODUCTION

Here we discuss a scheme that may be ideally suited to retrofit application to existing houses the south vertical faces of which receives much solar radiation even in midwinter. The scheme is much like that pioneered by S. R. Kenin and others of Solar Room Co., Box 1377, Taos, NM 87571. The ideas presented below were largely inspired by Kenin's work and publications. There are already several buildings that use his scheme, or schemes reasonably similar to it. An example is Egri House in Taos, NM. This house is described in detail in my book *Solar Heating Buildings of North America: 120 Outstanding Examples.*

PROPOSED SCHEME

A very-low-cost, transparent, walk-in enclosure, room, or greenhouse is installed along the south face of the given house. The enclosure may be, say, 24 ft. in east-west dimension, 8 ft. wide, and 8 ft. high. It may consist of two 0.006-inch sheets of tough plastic (Monsanto Film #602, for example) supported by curved tubular steel ribs spaced 18 in. apart. The two plastic sheets may be kept taut and 1 or 2 in. apart by pressurization of the air between them; the pressure is maintained by a tiny blower. The 2-inch layer of trapped air serves as insulation. Doors are installed in the end faces of the enclosure.

There are several openings in the vertical south wall of the house proper: some openings are near ground level, some are near ceiling level. The openings may be existing windows, or they may be specially made openings. On sunny days the enclosure receives an enormous amount of solar radiation, which heats the exposed (dark-colored) surfaces of the enclosure floor and north face and also the air in the enclosure. Room air circulates through the enclosure, carrying much heat from it into the rooms. The flow is maintained by gravity convection. At night, the above-mentioned openings are closed, so that the rooms of the house proper tend to stay warm and the enclosure is permitted to cool drastically. (Typically, it will not cool below about 40°F, and plants within the enclosure may be unhurt. In extreme conditions the cooling may be more drastic and may damage plants unless a little auxiliary heat is supplied.)

My understanding is that such an enclosure costs only about $1000 or $2000, if the house occupant is handy with tools and does most of the installing work himself. The materials, already cut to size and ready to assemble, are supplied by the manufacturer.

An enclosure of the size mentioned above has an effective aperture for radiation collection of about 300 ft^2. If the house in

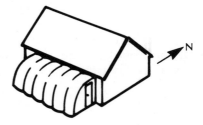

Fig. 1. Vertical cross section, looking west, of house and solar enclosure

Fig. 2. Perspective view showing enclosure supported by curved ribs.

question is of typical size and is in a typical location in USA, the enclosure may supply an important fraction of the winter's heat-need (50%, I guess).

It is hard to think of any other system that supplies so much heat (to an existing house) at such low cost. However, the enclosure will not be a great success if it is shaded by trees or neighbors' houses, or if prized shrubs occupy the space which the enclosure is to preempt. If the plastic sheets fail after one or two years of use and have to be replaced, the annual savings will be reduced; however, the sheets are so inexpensive that the net savings may still be large.

DISCUSSION

Why is it that use of the proposed enclosure is so much more cost-effective than using typical air-type collector panels mounted on the sloping roof? These reasons may be given:

1. Using an enclosure, the house occupant relies on the existing house for mechanical strength.

2. He uses the enclosed ground and enclosed vertical face of the house as absorbers. (If the house face is not dark in color, he can paint it a dark color.)

3. All of the installation work is done from the ground or stepladder. There is no need to climb up onto the roof; working high up on a roof can be time consuming and dangerous.

4. At any time the house occupant may walk into the enclosure to inspect or repair it. (He may be incapable of repairing a conventional collector panel situated on a roof.)

5. The shipping cost is low, because the plastic is shipped rolled up and the set of steel ribs is compact and lightweight.

   Other favorable considerations are:

The enclosure can be used as a greenhouse. Flowers and vegetables can be grown there throughout most of the year.

Moist air from the greenhouse will circulate into the house, providing a welcomed increase in humidity.

The enclosure may be used as a patio, sunroom, or playroom. Thus, it increases the useful floor area of the house.

In summer the enclosure can be dismantled and stored in a garage. Thus, the plastic sheets are not exposed to the most intense solar radiation and the highest ambient temperatures. Their useful life is prolonged.

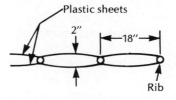

Fig. 3. Cross section of a portion of the enclosure showing two plastic sheets and three tubular ribs.

MODIFICATIONS

One could install a blower that would greatly speed up the circulation of room air through the solar enclosure and increase the total amount of heat delivered to the house proper.

The opening and closing of the ports between enclosure and rooms could be automated. Either lightweight, airflow-actuated dampers, or conventional electrically powered controls, could be used.

One could shorten the warm-up time of the enclosure and increase the amount of heat delivered to the rooms by making the enclosure virtually massless—by greatly reducing its dynamic thermal capacity. This can be done by spreading a 2-inch-thick layer of lightweight insulation on the floor and north wall of the enclosure and then installing a thin black sheet over the insulation. Then, practically no heat is delivered to the massive components of floor or wall; practically all of the heat is promptly transferred to the air. And since the thermal capacity of the 100 or 200 lb. of air in the room is equal to that of one fourth as great a mass of water (about 25 to 50 lb. of water), the air will heat up very rapidly. I estimate that its temperature will rise about 40 F. degrees in about two minutes, after the sun comes out from behind a heavy cloud cover. At the end of the day, little heat will be "left on base" in the collector floor or north wall and, accordingly, the enclosure will cool off very rapidly.

If much heat can be supplied by the solar enclosure to the rooms of the house, the rooms may become too hot by mid-afternoon. It might pay to install (in the house) a bin-of-stones to receive and store some of the heat.

OUTDOOR SLOPING AIR-TYPE THERMOSIPHON COLLECTOR BOX MOUNTED NEAR BASE OF SOUTH WINDOW

SUMMARY

Figure 1 shows the general design. At the start of a sunny day in January, a resident opens the lower sash of the window to which the collector is attached and extends the collector's septum-tongue 1 ft. into the room. Cold (60°F) air from the room travels south and downward along the lower plenum, then travels north and upward along the upper (sunlit) plenum, then (at 85°F) enters the room and rises toward the ceiling. The amount of energy delivered to the room is 5000 Btu (1.5 kWh) per hour or 25,000 Btu (7.3 kWh) per day. Late in the afternoon the resident telescopes the septum-tongue and closes the sash.

Cost of plywood, insulation, plastic film, weatherstripping, etc.: $100 (a guess).

Note that the collector operates automatically by gravity convection. The flow-rate adjusts itself automatically to changes in level of irradiation. No electric power is needed; normal operation continues even during failure of the electric supply. The dynamic thermal capacity of the collectors is so low that delivery of hot air can start within two minutes after the sun comes out from behind heavy cloud cover. The resident can close the sash at any time. If heavy clouds obscure the sun for a half-hour, say, with the sash left open, little room-energy is lost because (a) the sky itself delivers some radiation to the collector, (b) the collector is insulated on all sides, and (c) if the air in the collector becomes cold the flow ceases (the cold air in the collector is trapped there by virtue of its greater density). The device can be transported from factory to home on car-top or assembled at home by persons familiar with simple tools. It can be mounted in one hour by two such persons.

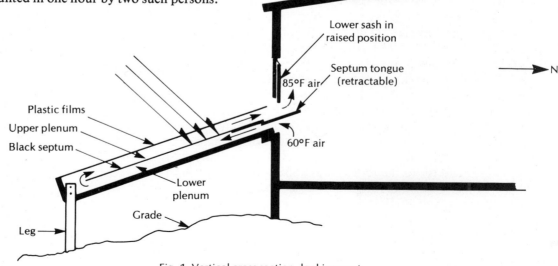

Fig. 1. Vertical cross section, looking west.

INTRODUCTION

The design was worked out in 1973 at the request of W. J. Jones of the MIT Energy Lab. He pointed out the great need for a cheap, easily constructed device that could be "bolted onto" an existing standard-type house to provide a moderate amount of solar heat to at least one room—to save fuel, save money, and provide a little warmth even in the event of total failure of fuel supply or electric supply.

The design employs a simple but effective principle that was demonstrated many years ago by S. C. Baer of Zomeworks Corp., PO Box 712, Albuquerque, NM 87103.

DESIGN

Figures 2 and 3 show many of the details of the proposed design. The collector is 4 ft. × 8 ft. in area and 1 ft. deep. The bottom, sides, septum, and spine are of ½-in. weatherproof plywood and are secured by screws. Most surfaces are insulated with ½ inch of polyurethane foam. The top consists of two spaced 0.004-in. films of weatherproof transparent plastic secured at the edges by means of pressure-sensitive tape. Insulator surfaces that receive radiation are black; the large horizontal black surface has vestigial ribs that encourage creation of some turbulence in the air flowing along there.

An adapter-panel at the upper end of the collector is secured to the outside face of the window frame by means of screws and is weather-stripped. The lower end of the collector is supported by two short legs (or two long poles, if the device is to be mounted at a second-story window).

PERFORMANCE

The nominal area of the collector is 4 ft. × 8 ft. = 32 ft^2; the effective area is 30 ft^2. On a sunny day in January the amount of solar radiation received per hour is (200 Btu/ft^2) (30 ft^2) = 6000 Btu, of which 5000 Btu (i.e., 1.5 kWh) is delivered to the room. The amount delivered during the day as a whole is (5 hr.) (5000 Btu/hr.) = 25,000 Btu (i.e., 7.3 kWh). This amount of energy is 4% of the total energy requirement of a well-insulated two-bedroom house in Boston on a typical 24-hr. day in January and is 7% of the amount of energy required of the furnace (much energy being provided also by miscellaneous sources such as stove, lights, appliances, human bodies, sunlight entering windows). On a sunny day, the energy delivered by the collector to the room in question would keep this room much warmer than other rooms—75 to 80°F as compared to 60°F, say. Any occupant who feels chilly can come to this room to warm up.

SAVINGS

If, on a sunny day in January, the collector cuts fuel consumption by 7%, it will cut it by larger amounts in milder months. On overcast days the device will not be used. I guess that, for the winter as a whole, the fuel saving will be 15%. If, in the winter of 1978–1979, the fuel cost for the house in question would be $600, the saving from use of the collector would be (15%) ($600) = $90. I would guess that if a manufacturer made and sold many devices of this type, his selling price would be about $250.

Thus in terms of the reduction in the amount paid for fuel, the device would pay for itself within 3 years. In terms of the reduction in worry concerning possible failure of the fuel supply or electric supply, it might justify itself in an even shorter period.

RECAPITULATION OF FAVORABLE FEATURES

See last paragraph of the Summary.

DRAWBACKS

Shrubs may be in the way of the device (but it may happen that the collector harmlessly "reaches over" the shrubs). Trees may shade the collector (but deciduous trees will produce little shade). Children, cats, squirrels, and birds may walk on the plastic films and damage them (but fences or chicken-wire covers could prevent this). Snow may stick on top of the device (but residents can scrape it off, and in any event, the sun will soon melt it off). Air-leaks at the window may occur unless the adapter is secured and weather-stripped carefully (but the leaks can be detected easily and plugged easily). The warm air produced is delivered to the ceiling (but this increases the amount of radiant heat from the ceiling; also, an electric fan can be used to circulate air from ceiling to floor). In a high wind the device might be ripped loose (but it would be easy to secure the legs to stakes driven into the ground or to a very heavy log). The expected life of the plastic might be only a few years (but replacing it would be simple).

MODIFICATIONS

(a) Install such devices on several first- and second-story windows.

(b) Employ a small electric fan to speed the flow of air along the collector and, thus, increase the amount of energy collected.

(c) Employ a special adapter which would permit mounting the device *vertically* below a second-story window.

(d) On the two sides of a sloping collector, install flaring aluminized "wings," to funnel additional amounts of radiation into the collector. Tie 12-in. diameter bundles of brush to the upper edges of the wings to dampen wind-gusts and reduce average wind-speed past the collector top and, thus, somewhat reduce heat-loss.

(e) Provide an alternative design that is longer and wider; use a fan.

(f) Provide an alternative design that is horizontal, runs parallel to the house, and uses a fan. Such a device can be very long. It benefits from radiation reflected from the white side of the house.

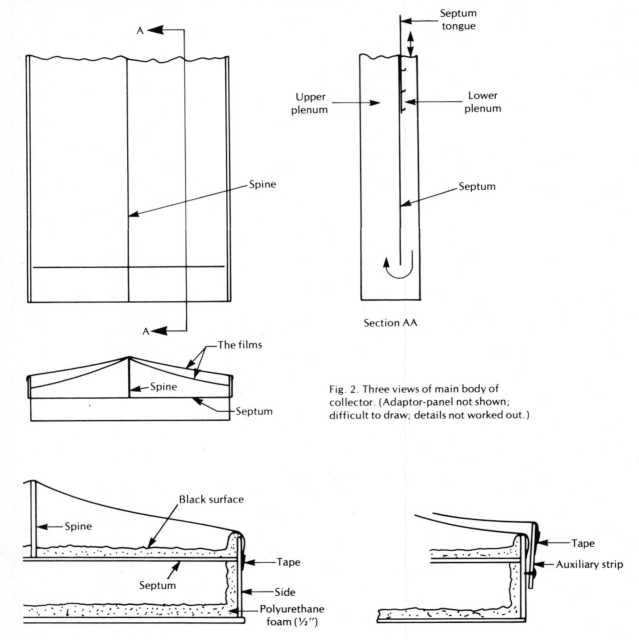

Fig. 2. Three views of main body of collector. (Adaptor-panel not shown; difficult to draw; details not worked out.)

Fig. 3. Diagrams show locations of insulation. Also they show the means of affixing the plastic films: first film is taped to collector side; then auxiliary strip is screwed on and second film is taped to it. The scheme provides an airspace between the two films. The spine increases the cross section of the upper plenum, keeps the films taut, and helps "slope" them to shed rain.

VARIATION PROPOSED AND TRIED OUT
BY W. SCOTT MORRIS

In 1977 W. Scott Morris of Box 4815, Santa Fe, NM 87501, tried out an "inverted geometry" scheme in which solar radiation is absorbed in the *lower* plenum. The septum is transparent; thus, the solar radiation passes through the (single) transparent cover, then through the septum, and is absorbed by a black sheet within the lower plenum. Preferably, this sheet includes many perforations (holes) and is mounted diagonally in the space available. Cool air from the room travels downward within the *upper* plenum and the warm air travels upward in the *lower* plenum.

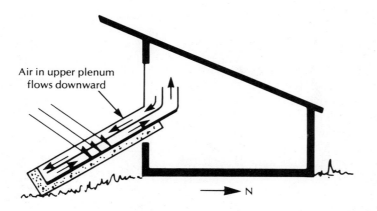

Fig. 4. Morris scheme employing inverted geometry.

The great advantage of this arrangement is that very little heat is lost by conduction or convection. The air that is very hot (the air in the lower plenum) loses little heat upward because the heat that leaks upward into the upper plenum is later carried down into the lower plenum and, thence, into the house; little heat is lost through the sides or bottom of the lower plenum because of the thick insulation used here.

A small drawback is that if a total of three transparent sheets are used, the reflection loss is larger than one might wish.

Some of Morris's designs have been described in the *Proceedings of the 2nd National Passive Solar Conference*, March 1978, p. 596.

INDOOR WATER-TYPE THERMOSIPHON SYSTEM EMPLOYING HINGED VERTICAL COLLECTOR-PANELS AND OVERHEAD CYLINDRICAL WATER-FILLED TANKS

PROPOSED SCHEME

Inside the room, close to the big, vertical, single-glazed south window, there are several vertical thin water-type collector panels that, by thermosiphon, serve a large overhead array of horizontal, cylindrical, water-filled tanks (of Kalwall Sun-Lite?). These are radiationally isolated from the room by manually controlled louvers which, when closed, constitute the ceiling.

The two tubes connecting a given panel to an overhead tank are flexible, and the panel can be rotated 90° (at start of day, for example, when room is cold) to let nearly all of the radiation penetrate deep into the room, warming it. Later, the panel may be turned so as to be parallel to the window, or so as to roughly face toward the afternoon sun, and collect much energy, which is automatically transferred to the top of the pertinent tank. The stratification that naturally occurs within the tank is helpful in that the water that flows to the panel (from the bottom of the tank) is cooler than the water in the upper part of the tank.

Each panel is 2 ft. wide, 7 ft. high, and is single glazed with cheap plastic (which does not have to withstand UV because it is protected by the big glass window). The backing includes 2 inches of insulation. Collection efficiency is high because the panel is in a warm environment. Note that most of the heat that leaks from the panel remains in the room and helps keep it warm. Typically, during the daytime, the panels are at about 45° from the big window; thus, they resemble a set of partially open louvers: they admit much daylight and allow a view toward the south and southeast.

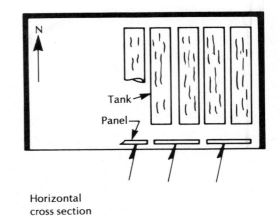

Horizontal cross section

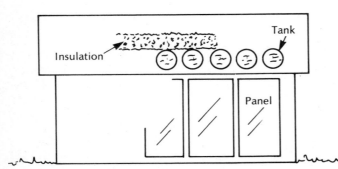

Cut-away view, looking north

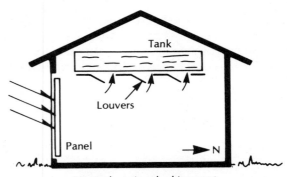

Vertical section, looking west
(piping not shown)

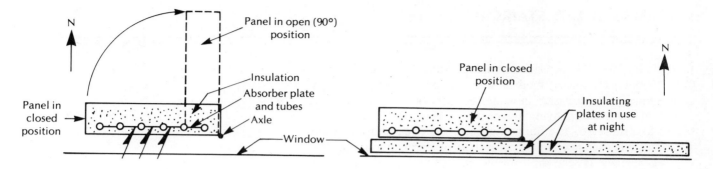

Horizontal cross sections

Because a large fraction of the solar energy entering the room is taken up by the panels and transferred to the tanks, the room does not overheat. At night, the ceiling louvers may be opened to permit radiant energy to flow downward from the tanks to the room. On cold nights the big window is insulated by means of big, 2-inch plates of rigid foam which clip into place. During the day these plates are parked on the insulated (back) faces of the panels.

The storage tanks, each 12 in. in diameter, are horizontal and run N–S. Each lies between the extra-strong ceiling joists. If each is 14 ft. long, it contains about 700 lb. of water. If there are 14 tanks in all, the total weight of the water is 10,000 lb.—equivalent (in thermal capacity) to 55,000 lb. of masonry. Because, with masonry, only the first 3 to 5 inches are actively used because the heat goes in and out with difficulty, and because the temperature swing in the masonry is only about 15 F deg., the 10,000 lb. of water is effectively equivalent to *several times* 55,000 lb. of masonry. At the end of winter the panels could be disconnected and stored in the garage.

MODIFICATIONS

Scheme S–220a

As above, except dispense with the insulating plates and use the panels themselves as the nighttime shutters. Make the supports for the panels compliant, so that the panels can be pressed tightly against the window at night.

Scheme S–220b

As above, except arrange to rotate the panels 180° at night, so that the panels' insulating *backing* presses against the window. The absorber sheet and its water-filled tubes will then remain at room temperature throughout the night, rather than becoming very cold.

Scheme S–220c

Install a very small blower that can circulate room air into the space surrounding the tanks. Thus, heat can be extracted from the tanks faster and more heat can be extracted.

In 1975 and 1976 N. B. Saunders of 15 Ellis Rd., Weston, MA 02193, invented a special kind of sloping, south-facing roof that admits much solar radiation in winter but very little in summer. As shown in the accompanying sketches, the roof has some resemblance to a staircase. The horizontal members (treads) are opaque and reflective; each may consist of a sheet of shiny aluminum. The vertical members (risers) are transparent and may consist of glass or tough plastic. Above the "staircase" there is a continuous, sloping transparent sheet that serves to shed rain and snow and reduce heat-loss through the roof assembly.

In winter the sun's rays that strike the treads are reflected toward the risers, pass through them, and enter the room. Also, some rays from the sun strike the risers directly; these rays also enter the room where they strike typical walls, furniture, etc.; or they may strike an especially massive wall installed with energy storage in mind. If heat is stored in this manner, the rooms remain warm throughout the evening without recourse to auxiliary heat.

In summer the sun's rays are more nearly vertical. Few such rays enter. Most are reflected steeply upward toward the sky. Thus, the room may remain fairly cool in summer.

The special roof provides, as an extra dividend, excellent daytime natural illumination throughout the room both in winter and in summer. Even deep within the room, little or no artificial illumination is needed during the day. The saving on electricity is considerable.

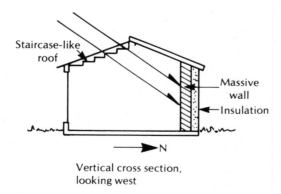

Vertical cross section,
looking west

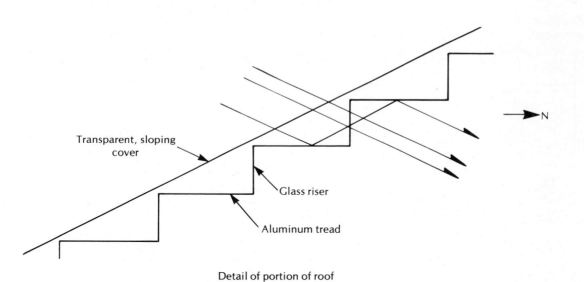

In winter, in one building (in Weston, Mass.) that has such a roof, the rays that pass through the roof strike a massive concrete wall 10 or 20 ft. farther north. In another building they strike some water-filled containers situated close beneath the roof. The containers are transparent, being made of Kalwall Sun-Lite; thus, besides storing much heat they allow much light to pass onward deep into the room to illuminate it. Controllable vanes close below the containers are used to block the flow of far-infrared radiation from containers to room when the room is already hot enough. (The buildings referred to are described in detail in my book, *Solar Heated Buildings of North America: 120 Outstanding Examples*.)

A drawback to the staircase-like roof is that much heat is lost through it on cold nights and cold days. The total amount of heat lost through it in midwinter months is only moderately exceeded by the heat *received* through it. The amount of heat lost could be reduced by installing, just below the roof, one or more sloping sheets of transparent plastic, preferably a kind of plastic that has low reflectance.

I believe that the roof in question is patented, or about to be patented, and I believe that the expression ''Solar Staircase'' has been trade-marked.

Transparent, sloping cover

Glass riser

Aluminum tread

N

Detail of portion of roof

STAIRCASE-LIKE ROOF ASSISTED BY THE COMBINATION OF REFLECTING-AND-INSULATING PLATES AND WATER-FILLED TANKS, ALL CLOSE BENEATH THE ROOF

Scheme S–200
5/12/78

SUMMARY

Figure 1 shows a scheme that somewhat resembles the Saunders staircase-like solar roof but may outperform it. The proposed roof includes insulation for the treads and also a set of thick insulating plates, which are mounted on hinges and can be closed at night—by pulling on strings—to insulate the roof as a whole. A set of water-filled containers is mounted just below the roof.

   In later paragraphs a great variety of alternative schemes are mentioned. Each has some advantages and some disadvantages.

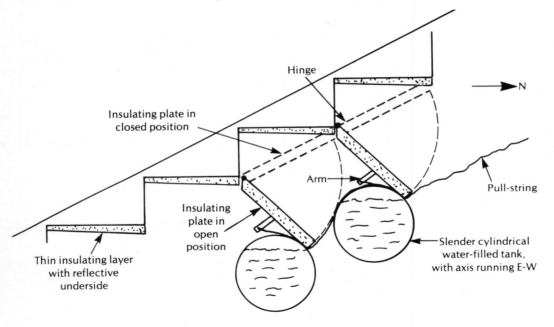

Hinge

Insulating plate in closed position

N

Arm

Pull-string

Insulating plate in open position

Thin insulating layer with reflective underside

Slender cylindrical water-filled tank, with axis running E-W

Fig. 1. Vertical cross section of a portion of the proposed system. The observer is looking west.

PROPOSED SCHEME

A large number of insulating plates are used—one for each tread-and-riser pair. Each plate is made of insulating material and is about 1½ in. thick. Each is hinged at its south edge. The hinge is attached to the north edge of the adjacent tread. The plate may be swung upward, so as to press upward against the next-farther-north tread-and-riser junction; thus, a considerable volume of air is sealed off. It is stagnant and, thus, has high R-value.

   Below each plate, not far from the south edge, there is a projecting arm that is 4 in. long which projects approximately downward. An elastic string is attached to the outer end of this arm. The string runs upward and north, toward the ridge of the roof. When the string is pulled, the arm is pulled, exerting a torque on the plate, and the plate swings upward to *closed* position. In winter the plate is closed in this manner at the end of each daytime. Pulling the string may be accomplished manually or by means of automatic equipment. For the many plates, there may be many long strings or one long master string to which many junior elastic strings are attached, one per plate. Because of the elasticity provided, each plate closes tightly irrespective of the other plates and irrespective of any small maladjustments of the lengths of the various strings.

In the morning, the strings are released and each plate swings downward under the influence of gravity, until it meets a mechanical stop.

The stops may consist of slender, horizontal, cylindrical, water-filled tanks with axes running E-W, suspended (from roof joists) fairly close below the staircase. These tanks receive much radiation that passes through the staircase, and they store much heat. Heat is transferred from them to the room by radiation, or with the aid of a fan or blower that picks up hot air from the uppermost part of the staircase region and delivers it to lower, cooler regions of the building. The Saunders scheme of controlling downward flow of radiant energy from the tanks may be used; it involves use of vanes.

Despite the presence of the tanks, much daylight penetrates deep into the room. Much radiation passes between the tanks. If the tanks are of transparent plastic and the water is kept reasonably transparent, much light may pass directly through the tanks and penetrate deep into the room.

Both faces of the insulating plates are made highly reflective. Accordingly, the plates, when in open position, help funnel radiation to the tanks. And in summer, when the plates are closed (raised), most of the solar radiation that reaches them is reflected up and out. A thin insulating layer, with reflective underside, is affixed to the lower surface of each of the treads.

DISCUSSION

The insulation system is fairly simple. The plates are hinged—no sliding motions are required. The plates open due to gravity. They are closed by the pull of a master string. No long push-or-pull rods are required. The hinges may be attached to the staircase proper or to the pertinent roof joists. The tanks are suspended from the roof joists, which should be made extra strong for this purpose. Each plate insulates one tread and one riser. Each plate, being reflective, tends, when open, to guide incoming radiation toward the tank this plate rests on; that is, the plates provide a small degree of concentration of the radiation. The special roof is somewhat expensive, but it entirely replaces the regular roof.

MODIFICATIONS

Scheme S-200a

Merely apply insulating material to the underside of each tread. This reduces heat-loss through the treads. If the underside of the insulating layer is made reflective, there is almost no decrease in the amount of radiation penetrating deep into the room. Preferably, the insulating slabs are thicker near the north edges than near the south edges.

Scheme S-200b

Double-glaze each riser. This reduces heat-loss here, but, unfortunately, reduces the transmittance of the staircase by about 10%.

Scheme S-200c

Install, below the staircase, a large transparent, sloping sheet. This, too, helps; but it reduces the transmittance. Also, dust, moisture, etc. may accumulate on this added sheet.

Scheme S-200d

Install hinged insulating plates that serve the risers only. To close the plates, one may employ strings or push-rods and control arms.

Scheme S-200e

Employ plates that serve tread-and-riser and are hinged at the *north* edges. To operate the plates, one may use string, push rods, arms, etc. Cantilevered weights may be used to make the plates (when released) hang with a slope that allows incoming radiation to pass by.

Scheme S-200f

Employ prism-shaped insulating blocks, hinged as shown, that fill entire spaces defined by treads and risers.

Scheme S-200g

Employ plates each of which is so large that it may serve, say, three adjacent pairs of tread-and-risers. This makes for fewer components, fewer strings. But the devices are heavier. Also they are so large that they may warp considerably.

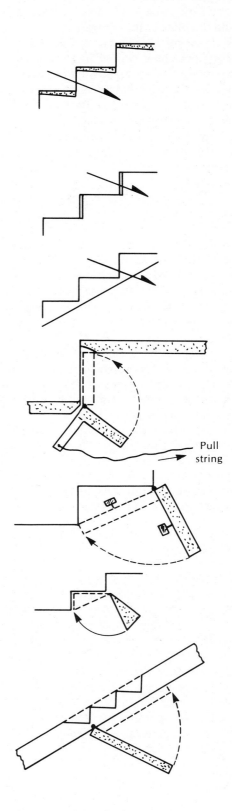

Pull string

Scheme S-200h

Employ large sliding insulating plates that may be slid parallel to the staircase. Provide good edge seals. To "open" the staircase, slide the huge plate toward a portion of the roof where there is no staircase. But sliding devices may stick. Also, providing tight seals for such devices is difficult. If wheels are needed to make the sliding force low, extra expense is involved.

Scheme S-200i

Employ, just below the staircase, a Beadwall window that has the same slope as the staircase. But the slope may be so gentle that the beads will not drain well. Anyway, the reflection loss is high and cost is high.

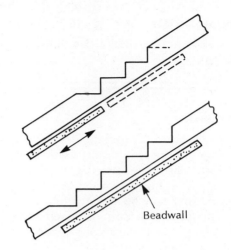

Beadwall

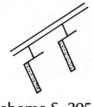

TRANSPARENT SLOPING ROOF EQUIPPED, ON UNDER SIDE, WITH SET OF INSULATING PLATES THAT MAY BE CLOSED BY PULLING ON ROPES

Scheme S-205
5/20/78

SUMMARY

Figure 1 shows the proposed solar roof. In includes two spaced, sloping, transparent sheets; below them are operable insulating shutters. The roof admits solar radiation during the daytime in winter and, at night, provides excellent insulation. During the summer the roof may be adjusted to exclude all radiation or to admit just enough to provide good illumination throughout the room.

The system is more complicated and more expensive than the Saunders staircase-like roof but is far superior to that system as regards conserving heat on winter nights.

ACKNOWLEDGMENT

The main features of the proposed system were suggested to me by John C. Gray.

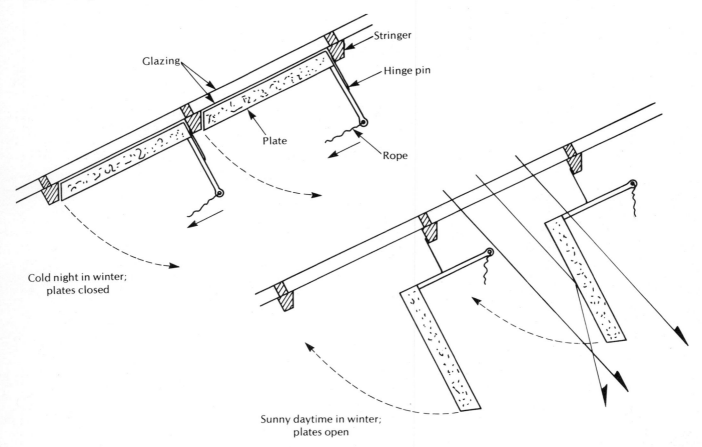

Cold night in winter; plates closed

Sunny daytime in winter; plates open

INTRODUCTION

The 1976 Saunders Solar Staircase roof provides excellent performance, admitting much solar radiation in winter and almost none in summer and has no moving parts. Its main drawback is that little insulation is provided—much heat is lost through the roof on cold winter nights. My report of 5/12/78 describes a scheme (S–200) that avoids this drawback. Operable insulating plates (shutters) are mounted just below the solar staircase roof.

On 5/12/78 John C. Gray pointed out to me that if one is planning to install operable shutters the solar staircase is not needed. The shutters, if properly designed, can do the whole job. This report indicates a specific design of shutter system that can do the whole job.

PROPOSED SCHEME

As the figures indicate, the sloping, south-facing roof is double-glazed with large sheets of transparent material that shed rain and snow. Immediately below the glazing there is a set of horizontal, east-west, wooden stringers and a set of insulating plates (operable shutters). Each plate is attached (near its upper or north edge) to a stringer. Each plate may be 2 in. thick (or 1 to 3 in. thick), may be 16 in. wide (or 12 to 24 in. wide) and 2 ft. long (or 1½ to 12 ft. long). Each consists mainly of lightweight insulating material and may include also a strong stiff sheet of plywood and a strong frame. The hinges are attached to the stiff sheet or frame.

The plates are closed by tension on ropes and, when the ropes are relaxed, are opened by the pull of gravity. When the plates are open, they hang sloping, aiming downward to the north. Thus, the incoming solar radiation, which also has a downward-to-the-north direction, passes readily between the plates and penetrates deep into the room, striking and warming the floor, north walls, etc. The reason that the plates hang sloping is that overhung, or cantilevered, hinges are used; see sketch.

Both faces of the individual plate have a highly reflective coating (white paint or aluminum foil). Thus, practically no radiation is absorbed by the plates and (even in summer or when oriented so as to block solar radiation) the plates do not become hot.

Each plate is controlled by a rope that is attached to an arm. The arm is simply an extension of the hinge; see sketch. Pulling downward-to-the-south on the rope exerts a positive closing torque on the shutter.

OPERATION

During sunny winter days the plates are left open. Solar radiation enters the room via almost the entire area of the roof and strikes and warms floor, walls, etc. Also, there is excellent illumination throughout the room. On winter nights and on very cold and heav-

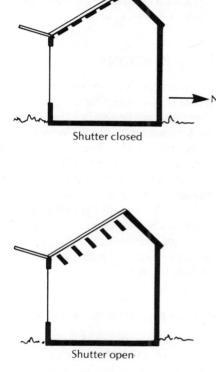

Shutter closed

Shutter open

ily overcast days, the plates are closed. Then very little heat is lost via the roof. During most of the spring and fall, the plates can be left open (or only partly open, if the rooms threaten to become too hot). On hot sunny days in summer, the plates are kept closed. They reflect solar radiation back to the sky. The rooms keep relatively cool. On summer days that are not very hot, the plates may be left partly open to provide good illumination. There is then no need to turn on electric lights; consequently, no heat is added to the room by such lights.

MINOR VARIATIONS

Stops may be provided to control how far the plates open, i.e., to insure that they all open to the same angle for the sake of appearance. Springs may be used to bias the plates toward open position or closed. In summer, special indents or stops may be used to limit the extent to which the plates are closed; for example, the occupant might want every plate to be 85% closed, so as to exclude nearly all of the solar radiation but admit just enough to provide adequate illumination. Instead of 85%-closing all plates, the occupant might completely close most of the plates and leave one or two wide open. Or, he could close all except two plates, and these two could be of special design—employing only *transparent* materials, such as a set of four high-transmittance plastic sheets as developed by Suntek Research Associates.

MAJOR VARIATION SCHEME S-205A

Install, close below the roof, some water-filled tanks. These would receive and store heat for later use, and they would reduce the tendency for the room to become too hot at the end of a sunny day.

FULLY CONTROLLABLE, 3-FT.-HIGH, INDOOR COLLECTION-AND-STORAGE WALL CONSISTING MAINLY OF SHALLOW, GLAUBER'S-SALT-FILLED BOXES

Scheme S-175
7/11/77

SUMMARY

Solar radiation that has passed through the large, double-glazed south windows immediately strikes a 3-ft.-high array of spaced, black boxes, each about 12 in. × 12 in. × ½ in. and each containing 3 lb. of a Glauber's salt formulation that has a melting point of 89°F and a latent heat of fusion of 108 Btu/lb. There are 1080 boxes and the total mass of salt formulation is 3200 lb. The total heat of fusion is 340,000 Btu. The radiation is absorbed by the boxes and the energy stored by them—inside an insulating housing—until needed to heat the rooms at night. The result is a high-performance passive solar heating system.

The system should perform much better than a system relying solely on storage in floors and walls, and somewhat better than a system using water-filled drums. It may be ideal for installation in existing houses that have large south windows but lack massive floors and walls. It is also ideal for use in second, third, and higher stories, where the enormous weight of a water-filled-drum system would require very heavy construction.

The present proposal becomes of interest only when and if suitable Glauber's-salt-filled boxes of proven durability and high performance become available at low cost.

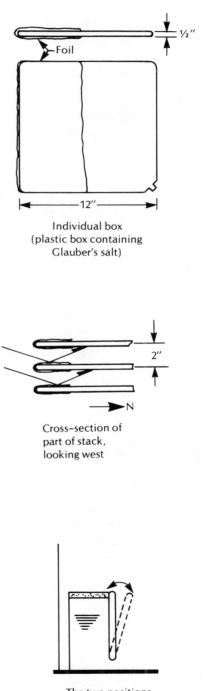

Individual box
(plastic box containing
Glauber's salt)

Cross-section of
part of stack,
looking west

The two positions
of the controller-plate

INTRODUCTION

Direct passive solar heating is reliable and cheap but performance is poor in these respects:

1. The heat imparted to the massive masonry floors and walls is not well controlled. On sunny days the rooms may become much too hot. On cold nights they become too cold. Floors and walls are hottest just when the rooms themselves are too hot, and coldest just when the rooms are cold.

2. The carrythrough is surprisingly small. It is small even if the total mass of floors and walls is, say, 200,000 lb. (100 tons). Why is it so small? One reason is that only a fraction of the masonry is effectively involved: (a) direct solar radiation strikes into the rooms only about 10 ft. typically; the total area of floors and walls struck is only about 10 ft. × 80 ft. = 800 ft.^{2}; (b) there is no simple way of getting the energy deep into the masonry quickly, i.e., during the six main sunny hours of a January day; if, in effect, only a 5-in.-thick layer of masonry is heavily involved in the heat-input process, the total mass involved is only 800 ft.^{2} × (5/12) ft.× 120 lb/ft.^{3} ≅ 40,000 lb. If the specific heat is 1/6 that of water, and a temperature

range of 12 F deg. is all that can be easily tolerated (remember: floors, etc. tend to be hottest when the rooms are already too hot and vice versa), the total amount of heat usefully stored is only $40,000 \times 12 \times 1/6 \cong 80,000$ Btu, i.e., only enough for about 5 to 8 hours on a cold January night.

3. The southmost 6–ft. region of the house must be kept relatively free of sofas, tables, bookcases, bureaus, etc., in order to permit direct radiation to penetrate deep into the rooms. Also, no large rug can be used here—it would greatly reduce the amount of heat entering the floor.

4. During sunny days the glare in the southmost 6–ft. region may be so great that the occupants shun this region.

5. The bright illumination of the southmost 6–ft. region destroys the privacy of that region with respect to strangers walking past the south side of house.

Note: Items 3, 4, and 5 imply that the southmost 6–ft. region is of greatly reduced usefulness. Say the usefulness is reduced by 50%. This is equivalent to getting *no* use from a 3–ft. region. Now, a 3–ft. region constitutes about 10% of the volume of the house, most houses being only about 30 ft. deep in all. If the house costs $60,000, a 10% non-usable portion represents a loss of about $6000, or perhaps $3000 if the space in question was simple and cheap to construct.

6. Use of an aluminum–faced outdoor porch (to considerably increase the amount of solar radiation collected) is ruled out because it would add greatly to the already–nearly–intolerable glare.

The good news is that most of these drawbacks would be considerably reduced (I guess) by the scheme described below.

AVAILABILITY OF BOXES

The present proposal is contingent on the availability of Glauber's-salt-filled boxes that cost not much more than $1 to $1.50 each, hold at least 3 lb. of Glauber's-salt formulation, furnish approximately the full nominal latent heat of fusion (about 108 Btu/lb.) and show no degradation of performance even after being frozen and melted a thousand times.

Boxes of about the desired size were produced in 1977 by Solar, Inc., of Mead, Nebraska. But I have little information on the performance or price. Larger boxes were being developed late in 1978 by Valmont Energy Systems, Inc., of Valley, Nebraska. Typical dimensions of box: 20 in. × 12 in. × 2 in. I lack information on performance and price.

(The general properties of Glauber's salt and other salt hydrates are discussed in Part 6.)

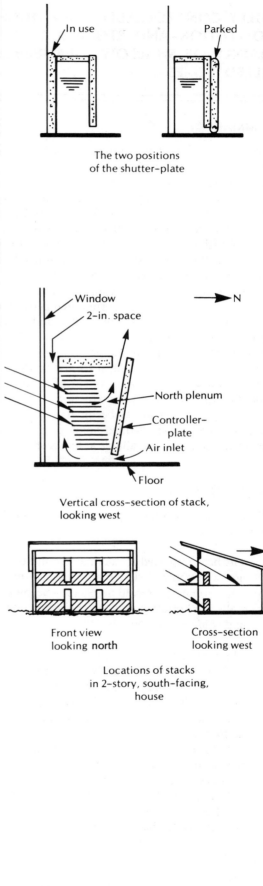

The two positions
of the shutter-plate

Vertical cross–section of stack,
looking west

Front view   Cross-section
looking north   looking west

Locations of stacks
in 2-story, south-facing,
house

PROPOSED SCHEME

The heart of the system, as applied to a new, two-story, well-insulated, south-facing house in New England, is a set of small (1 ft. × 1 ft. × ½ in.), sealed, black, near-horizontal, plastic boxes each of which contains 3-lb. of a formulation consisting mainly of Glauber's salt (sodium sulfate decahydrate, $Na_2SO_4 \cdot 1OH_2O$). The melting point is 89°F, the heat of fusion is about 108 Btu/lb., and the density is 1.5 times that of water. The boxes are arranged in vertical, 3-ft.-high stacks. Within each stack the boxes are spaced 2 in. apart on centers in order to (1) allow solar radiation to penetrate deeply into the stack and (2) provide ample room for gravity-convective flow of air. Each box is tilted 3 deg. downward toward the north to allow solar radiation to penetrate farther between boxes than if the boxes were horizontal. (The tilt must not be so great that the heavier components of the salt formulation gradually settle out more or less permanently.) The southmost half of each box is crudely wrapped with a single layer of aluminum foil to reflect the radiation deeper into the stack. Thus nearly all parts of each box receive radiation—even the wrapped parts receive energy, because the foil has about 10 or 20% absorption. The boxes are supported by a cheap, light-weight structure within the enclosing housing. The south face of the housing consists of a single sheet of cheap glazing (glass or plastic). The top of the housing is well insulated. The north face of the housing consists of a 2-in.-thick, insulating, *controller-plate* that is hinged at the bottom and can be swung outward (northward) at the top. Maximum swing: 7 deg. There is also a 2-in.-thick *shutter-plate* which can be manually (1) inserted in the 2-in. space between housing and window (at night, to prevent heat-loss from the south face of the housing and also prevent heat-loss through the pertinent area of the window), or (2) removed (e.g., during sunny day to allow solar radiation to penetrate into the stack). When not in use, the shutter-plate is parked, i.e., affixed (by loop and hook) to the north face of the controller-plate. Just below the controller-plate there is a permanently open slot, or vent. (The hot air within the housing tends to rise, hence never escapes via this slot.) Within the housing there are tapered plenum spaces on the south and north sides of the stack of boxes.

A total of 60 linear ft. of stacks is used, i.e., 30 linear ft. per story. The aggregate vertical south area of the stacks, i.e., the gross entrance aperture of the system for solar radiation, is 60 ft. × 3 ft. = 180 ft.²

The total number of boxes is: (6 per vertical ft.) (3 ft. height of stack) (60 linear ft. of stacks) = 1080.

The second story has a 5-ft. reflective porch which adds to the amount of radiation striking the second-story stacks. No glare problem is caused: the stacks normally block the occupants' view of the porch.

OPERATION

At the start of a sunny day in midwinter the occupant closes the controller-plates and parks the shutter-plates; solar radiation enters the stacks and the salt in the boxes melts. (What heats the rooms during the sunny day? The radiation that passes over the stacks and penetrates deep into the rooms.) At the end of the sunny day the occupant inserts the shutter-plates in the spaces between stacks and windows. Later, when the rooms need heat, the occupant opens the controller-plates to allow room air to circulate into the housings and pick up heat. In late spring and summer the sun's rays are so steep that they penetrate into the stacks very little. Fortunately, there is then little need to receive or store heat.

PERFORMANCE

How much latent heat does the system store? The total mass of salt hydrate is (1080 boxes) (3 lb./box) $\cong$ 3200 lb. Total amount of latent heat is thus (3200 lb.) (108 Btu/lb.) $\cong$ 340,000 Btu. (This provides about 4 times the carrythrough of the masonry system mentioned above.)

How much energy does the system receive on a sunny day in December? The answer is: about (1000 Btu/ft.^{2}) (180 ft.^{2}) (factor of, say, 1.2 to take into account the reflective porch) = 210,000 Btu. (The collection efficiency with respect to the radiation that has passed through the double-glazed south windows is very high; this is true even the first minute the sun comes out from behind a cloud and is true also even on cloudy days.)

> Note: The figure for *total* solar energy receipt by the *house as a whole* is much larger, inasmuch as there is much *direct* solar heating also, i.e., by radiation that passes just above the housings and penetrates deep into the rooms. Likewise the figure for *total storage* by the house as a whole is much larger, inasmuch as floor and wall areas receiving direct solar radiation store some energy.

Total area of box surface giving out heat when air circulates through system: (2 ft.^{2} per box) (1080 boxes) = 2160 ft.^{2} Gratifyingly large.

My guess is that the heat supplied by the overall system described above (heat from salt hydrate system and heat from floors, etc.) would constitute about 70% of the winter heat-need of the house under discussion.

OTHER GOOD FEATURES

All south rooms receive much daylight—via the upper portions of the big south windows. Only the lower 3-ft. portions are blocked by the special housings.

Good view toward the south is available in all south rooms—the occupants merely look over the special housings.

All of the big drawbacks of ordinarily passive systems are greatly reduced.

The tops of the housings are fixed. Accordingly the occupants can lay books, magazines, potted plants, etc., on them.

The housings are small and low enough so that an adult can lean over them and peer down behind them, or reach down behind them. Also he can easily reach and operate roll-down window shades and can reach the windows to wash them.

The boxes are easily accessible. The controller-plates can be removed entirely, fairly simply, as when the occupant wishes to inspect the boxes or wishes to remove them or replace them.

Domestic hot water can be preheated in pipes within the housings.

Auxiliary heat can be applied by off-peak-powered electric heating strips running along the bottoms of the housings.

In summer the shutter-plate can be operated in reverse to exclude radiation.

MODIFICATIONS

Apply the system to a one-story house, multi-story house, office building, etc.

Retrofit the system to an existing house that has big south windows but lacks massive floors. The system may be near-ideal for retrofit application.

Use a salt hydrate that melts at 120°F (e.g., "hypo") instead of 89°F. Or use some of each: use a 120°F formulation for heating domestic hot water and an 89°F formulation for space heating.

Use stacks of varying heights. Or use 7-ft.-high stacks in some places and no stacks at all in other places.

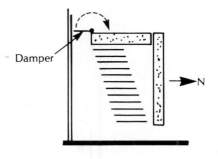

Damper

Use, at lower parts of south windows, single (not double) glazing. This may suffice inasmuch as these windows are protected, most of the time, by shutter plates.

Leave off the thin glazing on the south side of the stack and install a convection-stopping damper at top of 2-in. space between stack and window. This would reduce reflection loss and cut cost. (I am indepted to J. C. Gray for suggesting this improvement.)

Automate the operation of the controller-plates and shutter-plates. Or use Beadwall or Skylids.

Use small fans to coax heat out of housings faster.

Use an array of water-filled drums on first story, where great weight is easily supported, and use the (much lighter) salt hydrate system on upper stories.

COMPARISON WITH SET OF WATER-FILLED DRUMS

Water-filled drums provide, of course, excellent performance at very low cost. Drums are cheap. Water is almost free. Automatic gravity convection of the water inside the drums is helpful. Water can help keep rooms cool in summer besides helping keep them warm in winter.

But the salt hydrate has, I believe, better performance. The proposed system stores about 3.5 times as much heat per pound as water used over a 24-F-degree range and stores about 1.5 times as much heat per cubic foot of material-and-voids (assuming that the drum system is 30% voids and the salt system, with its well spaced boxes, is 75% voids). It has about twice the surface area for heat output, per unit volume of assembled system. Corrosion (of the plastic boxes) should never occur. Freeze-up-produced bursting cannot occur inasmuch as the salt hydrate shrinks when it freezes. No antifreeze is needed. Every component is lightweight; the house occupant could transfer the entire salt hydrate system to a garage in summer, to make the house roomier.

SYSTEM EMPLOYING WATER-FILLED BAGS ON TWO-LEVEL ROOF EQUIPPED WITH HINGED PLATES THAT REFLECT, INSULATE, AND ALSO SHED SNOW

SUMMARY

H. R. Hay's Skytherm House at Atascadero, California, is a one-story house. It employs water-filled plastic bags on the ceiling, and the bags are covered at night by insulating panels that run on horizontal rails. The house performs excellently—in California. It is presumably *not* suited for use in New England.

I propose a variation that *is* suited for use in New England. It is a two-story house; thus its volume-to-surface ratio is more favorable. Employing several large, crude mirrors, it intercepts 2½ times as much direct solar radiation (in New England in January) as the Atascadero House intercepts; that is, it has 2½ times the gross aperture. The "hot aperture", however, is much smaller; thus the steady energy *loss* (whenever the panels are open) is smaller. The panels are swung on horizontal hinges; tilting them upward tends to dump off snow; their undersides are crude mirrors. I guess that this system would provide near-100% solar heating.

In 1975 J. Hammond built a house having some resemblance to the one proposed here.

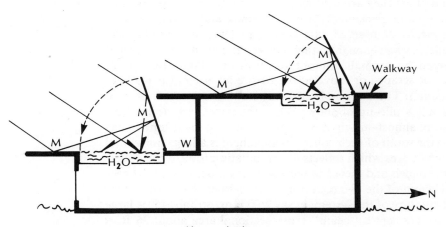

Cross section of proposed house, looking west.
Note the two water-bag areas (H_2O) and four mirror-like areas (M).

INTRODUCTION

Harold R. Hay's Skytherm House at Atascadero, California, was economical to build and has performed excellently, according to the evaluation reports by K. Haggard et. al. of California Polytechnic State University at San Luis Obispo. It achieves 100% solar heating and, in summer, near-100% solar cooling.

But it was not designed for use in a region (such as New England) where, in winter, outdoor temperatures are much lower, the sun's rays are more nearly horizontal, and heavy snowfalls occasionally occur.

Can Hay's scheme be modified so as to be suitable for New England? I think so.

PROPOSED SCHEME

This hypothetical 40 ft. × 40 ft. house, located, say near Boston, has two stories and is superbly insulated with the equivalent of 8 in. of fiberglass. There are two collection areas: at south portion of first-story roof and at north portion of second-story roof. Each collection area employs water-filled plastic bags each of which covers an area 36 ft. × 8 ft.; the thickness of water layer is 16 in. Above the water bags are two transparent plastic sheets (held apart by air pressure) that provide top insulation. The water bags rest on a black plastic sheet which in turn rests on a thin metallic ceiling of the room below, heating it by radiation, etc. To the north of each water-bag there is a 50-ft.-long, 8-ft.-high array of 4-ft.-wide panels that are hinged at the lower edges and slope 70° from the horizontal. The undersides of the panels are crude mirrors and reflect much solar radiation toward the water-bags. The panels include 3 inches of polyurethane-foam insulation. At night the panels are closed to prevent loss of heat from the water-bags. When open, the panels are held rigidly at the prescribed angle by ropes and clips; they are held against sturdy beams or posts. The panels are counterweighted for easy opening and closing—manually, by means of braided wire cables, pulleys, handcranks, etc. Each panel is small enough so that it can be moved manually even if encumbered by a small amount of snow. (A large amount of snow can be scraped or shoveled off before an attempt is made to tilt the panel.) There is ample space in front and in back of each panel to allow pile-up of snow. When the panels are *down*, they comprise an almost-air-tight enclosure, thanks to inclusion of edge seals. To the south of each collector-area there is a large horizontal crude-mirror area which reflects direct radiation toward the underside of the panels and thence to the water-bags. Because of the geometrical layout of the two collector areas, neither shades the other. The full aperture of the system, at say 2:00 p.m. on Jan. 15, is large enough to intercept a "beam" cross-sectional area about 36 ft. × 30 ft., i.e., about 1000 ft.^{2} measured in a plane *normal* to the sun's rays. The area of the "hot aperture" is only (2) (8) (36) = about 600 ft.^{2}

I guess that this system would provide near-100% solar heating of the proposed house.

COMPARISON WITH SKYTHERM HOUSE

At a typical hour on a typical midwinter day the proposed system intercepts about 2½ times as much direct radiation as the Skytherm

House, thanks to the staggered arrangement of the collectors and the use of crude mirrors. The volume of water used is about the same; in the proposed scheme the area of water-bags is smaller but the depth of water is greater. The temperature of the water may be higher, thanks to (a) the "double airbag" above the water-bags, and (b) the large ratio of cold aperture to hot aperture. The area radiating heat to the rooms is smaller than in the Skytherm House; thus the heating is not quite so uniform or rapid. The problem of accumulation of snow, and interference (by snow) to the operation of covers, is largely solved.

Some *troubles* with the proposed system are: (1) the large numbers of ropes, etc., needed to operate the panels (18 panels in all) and the nuisance of moving the panels approximately twice a day, on the average, all winter; (2) the imperfect reflectivity of the crude mirrors—reflectively of about 75%; (3) the large mass of water to be supported high above ground; (4) the marginally adequate heating of north part of ground floor.

CONCLUSION

The proposed system should work reasonably well in New England. Its cost may be acceptably low.

Designs by Others

In about 1975 Jonathan Hammond built a house employing a solar heating system somewhat similar to the one proposed here. The reflector is opened and closed by means of a hydraulic actuator. See *Passive Solar Heating and Cooling, Conference and Workshop Proceedings. May 18-19, 1976, Albuquerque, NM,* Report LA-6637-C by Los Alamos Scientific Laboratory; p. 153.

In about 1977 the Farallones Institute built a small house called Cabin B, with a somewhat similar solar heating system. See *Proceedings of the 2nd National Passive Conference,* Vol. 2, p. 298. Also *Solar Age,* July 1978, p. 20.

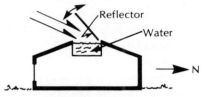

Hammond House: Cross section, looking west

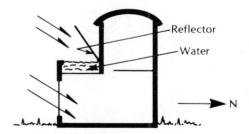

Farallones Cabin B.; Cross section, looking west

THE BUCKLEY TWO-COMPARTMENT, THERMOSIPHON, NON-REVERSING, SOLAR-ENERGY-ABSORBING-AND-STORING, WATER-FILLED TANK

INTRODUCTION

Shortly before 1974 Shawn Buckley of Massachusetts Institute of Technology invented a special kind of tank designed to be incorporated in the south wall of a house (see the accompanying figure) and to absorb solar radiation on sunny days, store the absorbed energy, release the energy to the house-interior during the night, and *not* lose energy to the outdoors at night.

Such a tank has two compartments, called S and N, comprising the south and north parts of the tank. Both contain water. Compartment S, which receives the solar radiation and becomes especially hot, is thin; it contains only a small amount of water. Glazing may be applied to its south face. Compartment N is much thicker—10 or 20 times thicker—and contains a very large quantity of water. Between the two compartments there is a vertical insulating partition, or septum. There are openings through the septum at top and bottom which permit flow of liquid from one compartment to the other. Associated with the upper opening there is a special valve, employing oil, that greatly improves performance, as explained in a later paragraph. Appropriate faces of the assembly are thermally insulated.

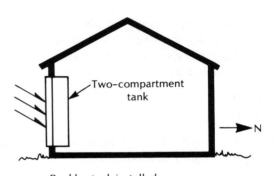

Buckley tank installed in south wall of house. Vertical cross section, looking west.

HOW THE VALVE WORKS

To facilitate the account of how the system operates and to make it clear why the special valve is so helpful, we consider first a simple device, shown in Fig. 2, that has no special valve and is incapable of performing properly. The only liquid used is water. The tank has been filled just full enough so that, ordinarily, the septum projects about ½ inch above the level of the water; thus no water can flow over the top.

When the sun comes out and heats the water in compartment S, the level of the surface of the water here tends to rise—see Fig. 3—because water, when heated, expands, i.e., becomes less dense (more buoyant). But it does not rise enough to permit flow of water over the top of the septum. Thus no solar energy received by compartment S can be transferred to compartment N.

If the compartments had initially been filled to a higher level, flow over the top of the septum might have occurred on a sunny day; but unfortunately a reverse flow might have occurred under some circumstances and much heat might have been lost to the outdoors.

Now we consider a tank that has a check valve that employs oil (mineral oil, e.g.) and we show how this valve facilitates circulation

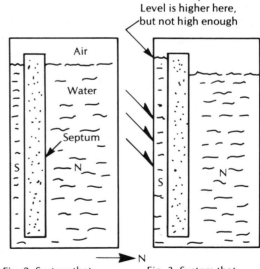

Fig. 2. System that has no valve. Nighttime.

Fig. 3. System that has no valve. Sunny day.

of hot water from S to N and yet prevents reverse flow on cold nights or cold overcast days. We consider an over–simplified design of valve, shown in Fig. 4. The oil, which does not mix with water, is assumed to have a density 0.9 times that of water; thus the oil permanently floats on the water. The oil layer is, say, 2 inches deep, and extends well above the top of the septum.

The interesting fact is that if the water in compartment S is heated to high enough temperature that, in the absence of any oil, the surface would rise by 0.2 inches, when the oil is present the surface tends to rise by about ten times this amount, i.e., about 2 inches. But this is more than ample to allow the water to pass over the top of the septum into compartment N. See Fig. 5. In summary, the overlying layer of oil acts as a height–change amplifier, or more exactly an amplifier of change in level of the surface of the water. The closer the density of the oil is to the density of water, the greater the amplification. In practice, the amplification is great enough so that flow of water, and transfer of heat, from S to N commences as soon as S is 1 or 2 F degrees hotter than N. The more intense the solar radiation and the greater the temperature difference between compartments S and N, the faster the flow.

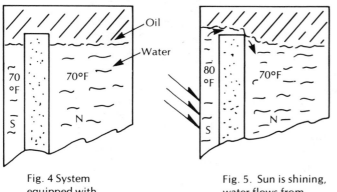

Fig. 4 System equipped with oil–type valve. System is in equilibrium.

Fig. 5. Sun is shining, water flows from S to N over the top of the septum.

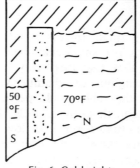

Fig. 6. Cold night. No reverse flow occurs.

WHAT HAPPENS ON A COLD NIGHT?

On a cold night compartment S, being very close to the outdoors, cools off rapidly. Does one expect, then, that water will flow (over the top of the septum) from N to S?

No. The valve (that is, the combination of the oil and the top of the septum) prevents this. It prevents it because of a crucial asymmetry: a crucial inequality of areas. In compartment N the water-vs.-oil interface area is very large while in compartment S the interface area is very small. Accordingly, even if the level of the water in S were to fall 1 inch, the rise in level of the water in N would be only a small fraction of an inch—not enough to permit water to flow over the top of the septum.

Notice that the oil continues to act as a height-change amplifier, and it amplifies height changes in both compartments. But because the interface area of N is much greater than that of S, the height change in N is negligible even after amplification. (Ten times practically nothing is still practically nothing.) Thus no water flows over the top of the septum from N to S.

In summary, it is very easy for a large water–height–change to occur is S but a large change cannot occur in N. Thus when the sun shines water can flow over the top of the septum from S to N, but on a cold night the reverse flow cannot occur. The oil valve does its job excellently—it permits one–way flow only, and even this flow does not occur except when it is advantageous, i.e., when S is hotter than N.

Buckley calls the system a *thermal diode* because it performs analogously to the common kind of electronic diode tube, which permits only one–way flow of electrons. The name seems to me not fully suitable since (a) some diodes do not have this property, (b) some triodes, pentodes, etc. *do* have the property, and (c) a diode is a kind of valve, but Buckley's device is mainly a storage system— the valve representing only about 1% of the volume and 1% of the cost.

SMALL SIZE OF VALVE

In practice, the oil valve is small. Instead of occupying the entire top of the tank, it consists of a small chamber (which retains the key feature of having two areas that are of very different size). And the water, instead of flowing into the chamber via one of its sides (which would mean that the valve would perform poorly if the assembly were slightly tilted), flows in centrally via a central vertical tube. Also, the valve is situated somewhat lower down, so that it will perform properly even if the tank has been slightly underfilled. Fig. 7 indicates, schematically, the location and design of a small valve with central tube.

ACTUAL DIMENSIONS

In practice the tank may be 8 ft. high, 4 ft. wide, and about 10 in. thick overall. Compartment S is extremely thin, to minimize warm-up time and minimize the amount of energy lost from it at the end of a sunny day. Usually the south face is glazed (and, in summer, the space between glazing and compartment S may be vented to make sure that the water in S will never get so hot that it boils.) The tank faces that are not intended to receive or distribute energy are well insulated. Distribution of heat may be facilitated by incorporation of air-ducts within or immediately adjacent to compartment N. The overall weight of the panel, filled, is about 600 lb.

Is Antifreeze Needed?

In some types of Buckley tank, compliant (deformable) elements are incorporated in the north wall of compartment S, with the con-

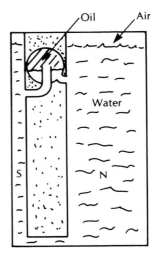

Fig. 7. Small oil valve with central vertical tube. Drawing not to scale.

sequence that freezing of the water in this compartment does no damage. Of course, antifreeze could be used if appropriate precautions were taken.

Compartment N requires no special precautions. It is almost inconceivable that the water in this large indoor compartment would freeze.

Application to Heating a House

If a great many of the Buckley tanks were incorporated in the south wall of a building, or if a quantity of (modified) devices were incorporated in the south-sloping roof of the building, a large fraction of the winter's heat need might be supplied by the sun.

The system could easily be modified to provide heat for domestic hot water.

Status of Project

Late in 1978 efforts to arrange for commercial production of the Buckley device were continuing.

A Final Question

One might ask this question: Inasmuch as the tank includes two compartments and each must be insulated from the other, why not use *two separate tanks?* One of them (the thin one) could be mounted on the vertical south wall of the house and the other (the thick one could be mounted somewhat higher up, close to the ceiling, or just above the ceiling. Then, because the larger tank is higher up, no valve at all would be needed. Gravity convection, via connecting pipes, would do the entire control job automatically. Also, less prime space would be taken up and leakage of heat from the big compartment to the small one would be entirely avoided. Admittedly, there are impressive advantages in having the two tanks teamed together as one integral unit; but there are impressive advantages the other way also; if two separate tanks are used, installation and maintenance may in some situations be easier and there is much greater freedom of choice as to where to place the larger portion of the equipment.

A big variation would be to place the larger tank in the basement and employ a small pump to circulate water from the smaller tank (on vertical south wall or on the sloping south roof) to the basement tank. But this is a conventional active, water-type solar heating system!

Thus the question is: Why, or in what situations, does one wish to combine the thin absorbing collector and the thick storage system into a single assembly?

References

The Buckley solar-energy-absorbing-and-storing device is described in *Solar Energy Digest*, Jan. 1977, in *Solar Energy 20*, 495 (1978), in *Solar Age*, p. 22 (April 1978), in *Proceedings of the 2nd National Passive Solar Conference,* Vol. 2, pp. 271 and 469 (1978), and in many special reports.

TWO-COMPARTMENT, THERMOSIPHON, SOLAR-ENERGY-ABSORBING-AND-STORING, WATER-FILLED TANK WITH FLOATING OUTLET THAT PREVENTS REVERSE FLOW

Scheme S-48
9/28/74

INTRODUCTION

The scheme proposed here is much like Buckley's scheme and performs approximately similarly. However, a very different kind of valve is used—a floating valve that is simple and performs well in a great many respects. No oil is used. The proposed system is less sensitive than Buckley's, but may be superior in some ways.

PROPOSED SCHEME

Figure 1 shows a simple embodiment of the proposed scheme. The main component is a rectangular, 4-ft.-high tank of galvanized iron. It is placed so as to form part of the south wall of a house. The south face of the tank is black on the outside; on sunny days it absorbs much solar radiation. The north face of the tank is either bare or insulated: bare when the occupants want the tank to deliver heat to the rooms, and covered with a 2-in.-thick insulating plate at other times. The other four faces are permanently insulated.

The tank is about 95% filled with water, and a vertical, 2-in.-thick insulating septum divides the tank interior into two compartments (south, S; north, N) the thicknesses of which are in the ratio of 1-to-30. The septum is sealed to the tank sides and top so that no water can pass through it except via a horizontal slot at the bottom and via a ½-in.-dia. hole about 40 in. above the bottom. At the top of the septum there is a notch, or "saw-cut", to allow air-pressure equalization in the tops of the two compartments.

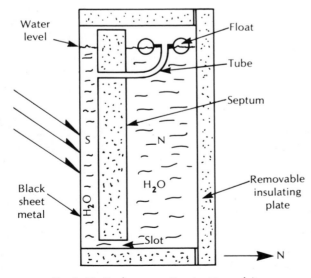

Fig. 1. Vertical cross section (not to scale)

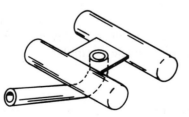

Fig. 2. Float and tube, perspective view.

There is a float on the water in Compartment N. It supports the north end of a flexible, ½-in.-dia. tube and holds this (open) end 0.01 in. above the level of the waterline of the float. The S end of the tube engages the hole in the septum. See Fig. 2.

The tank top is separate; it can be lifted off; it is not hermetically sealed.

OPERATION

If the tempertures T_S and T_N of the south and north compartments are the same, the water densities here are the same, and accordingly the water levels are the same. Nothing tends to make the liquid level in the north end of the tube higher than the float waterline, and no liquid emerges from this end of the tube. In summary, there is no circulation of liquid. See Fig. 3.

If the weather is cold and overcast, the temperature of the south compartment slowly decreases, the density here increases (height of water column decreases), but the total mass of water in this compartment remains constant. Thus, there is no tendency for any flow to occur through the slot or through the tube. In summary, there is no circulation of liquid. See Fig. 4.

If intense sunlight strikes the south face of the tank, the water in the south compartment heats up and expands (the density decreases). And because the mass of water here remains unchanged (initially), the height of the water column increases, the pressure within the tube increases, the liquid in the north end of the tube tends to rise—and after it has risen more than 0.01 in. it begins to overflow into the north compartment as a whole. Thus water is transferred from S to N. Whereupon, water at the bottom of the tank begins to flow, via the slot, in the opposite direction: from N to S. In summary, clockwise circulation starts. See Fig. 5.

In conclusion: during the sunny daytime the system accepts and stores energy, and on cold nights it refrains from losing energy to the outdoors. Thus it is an effective solar-energy-collection-and-storage system.

Fig. 3 $T_S = T_N = 70°F$

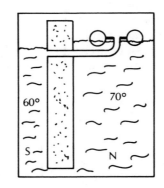

Fig. 4 $T_S < T_N$

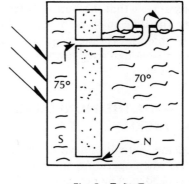

Fig. 5 $T_S > T_N$

SENSITIVITY

At the start of a sunny day in winter, as T_S slowly increases and exceeds T_N, when does the (desired) clockwise flow start? It starts when T_S exceeds T_N by about $2\frac{1}{2}°F$. Water at $72\frac{1}{2}°F$ is less dense by 0.03% than water at 70°F; hence the heights of liquid in the two compartments differ by 0.03%. Note that 0.03% of the 40-in. nominal height is 0.012 in.—which slightly exceeds the height increment built into the float-and-tube-end system. Therefore flow of liquid starts.

DISCUSSION

The system is very stable. Its sensitivity remains unaffected by modest-size changes of many kinds. For example, the sensitivity remains the same even if

- a little too much (or too little) water has been put into the tank (the float retains the same crucial increment in relative height),

- some water gradually evaporates, lowering the water levels,

- the tank sides gradually bulge more and more or are dented in by some children playing nearby,

- the septum slowly becomes warped, compressed, or soaks up some water,

- the tank is mounted with a slight slant in any direction,

- a 1/8-in. layer of sediment accumulates on the bottom of the tank,

- a liter of water is ladled out, to fill a hot-water bottle, say.

It is essential, however, that the buoyancy and mass of the float remain constant, so that the built-in increment in height will remain constant. The float must be made of dimensionally stable, water-impervious material (brass, glass, or certain plastics). The tube must be flexible enough so as always to exert the same downward force on the float.

COMPARISON WITH THE PIONEERING
SCHEME BY S. BUCKLEY

The Buckley scheme, with its oil valve, has order-of-magnitude greater sensitivity, when operating normally; but I'm not sure that such sensitivity is needed. The proposed floating outlet provides, I think, fully adequate sensitivity. If the Buckley oil valve is expensive (which I expect it is not) or fails to be highly durable, the scheme proposed here may deserve attention. It has been tried out by S. C. Baer and, I understand, performed well.

MODIFICATIONS

Summer-vs.-Winter Reversible Scheme

Use an especially slender float and mount it in the *south* compartment. Connect the crucial control tube to this float. This makes the system operate in opposite manner: circulation occurs only when the *north* compartment is the hotter one. Such operation is desired in summer—to *cool* the main quantity of water.

Various alternative schemes could be used for reversing the operation. For example, the float could remain in the N compartment and could be connected to a tube running from a hole near the *bottom* of the septum (the slot there would be closed, and the hole near the top of the septum would be left open).

Putting the Stored Hot Water to Additional Uses

One could use the large compartment full of hot water in conjunction with the domestic hot water system, to provide solar-preheating of this water. One could install a faucet near the top of the large compartment so that persons could extract small quantities of hot water for preparing a cup of hot cocoa, for filling a hot water bottle, for washing their hands, etc. At the same time, a cold-water inlet pipe would be installed near the bottom of this tank and a conventional float-valve would be employed to keep the tank filled to approximately a constant extent. Occasional small changes in liquid level and occasional addition or removal of a small amount of water would not interfere with the collection and storage of solar energy.

SET OF SOLAR-ENERGY-ABSORBING-AND-STORING WATER-FILLED TANKS THAT ARE SPECIALLY SHAPED AND TILTED SO AS TO PREVENT REVERSE FLOW

Scheme S-46
9/14/74

INTRODUCTION

Here is another collection-and-storage device that performs in somewhat the same way that the Buckley tank performs.

PROPOSED SCHEME

Install, as part of the south wall of the building, a set of long, thin, tilted, oblique-ended, water-filled tanks the south ends of which are painted black. As indicated in the first sketch, each tank tilts upward to the north. Also, within each tank and fairly close to the south end of it, there is an insulating plate which, in effect, divides the tank into a very small compartment and a very large compartment. Water can circulate from one compartment to the other via passages just above and just below the insulating plate. Removable insulation is provided on the outside of the tank sides and north end; when the insulation is removed, heat can flow from tank to room.

While the sun shines, the water in the small (south) compartment is heated, and circulates by gravity convection to the large compartment. During cold nights, almost no reverse circulation occurs because most of the water in the large compartment is higher up than the small compartment.

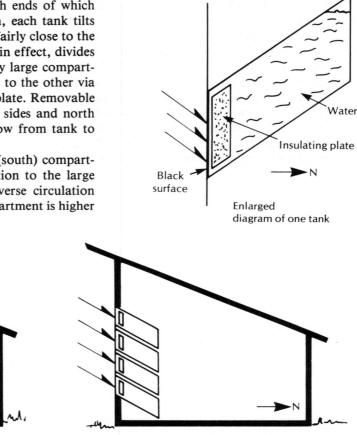

Water

Insulating plate

Black surface

N

Enlarged diagram of one tank

Set of four tanks

N

Vertical cross section, looking west

N

Tanks oriented to reject heat in summer

To convert for cooling in summer, merely reverse the slope by rotating each tank 180° about a horizontal north-south axis. Then the hot water tends to accumulate in the tank ends nearest the outdoors (because these are the higher ends), maximizing the rate of loss of energy to the outdoors and minimizing the amount of energy transferred to the large compartments.

MODIFICATIONS

Scheme S–46a

Change the shape of the individual tank. Instead of using a rhombic shape, use a shape consisting of a thin hollow plate (the small compartment) and a horizontal east–west cylinder (the large compartment), the latter being mounted at the upper end of the former. The cylinder may be just above head–height, and is thus entirely out of the way. A short pipe allows flow of water from the top of the small compartment to the top of the large compartment and another pipe allows flow from the bottom of the large compartment to the bottom of the small compartment. The pipes are not shown in the sketch.

In summer the equipment is inverted to insure one–way circulation that helps keep the room cool.

To speed the transfer of heat from the cylindrical tank to the room (in winter) or vice versa (in summer), a small fan could be used to direct a stream of room air at this tank.

Scheme S–46b

In a system that is intended for use in winter only, the thin compartment could be mounted outdoors, i.e., close against the vertical south wall. Thus the need for special wall construction is avoided, and the system could be applied to existing buildings.

The outdoor–mounted compartment could be tilted at, say, 60° from the horizontal, so as to receive more solar radiation.

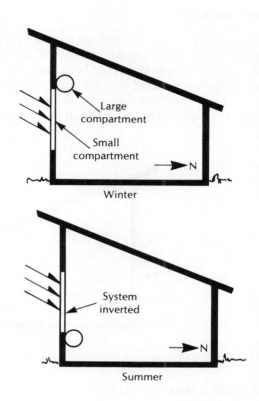

Winter

Summer

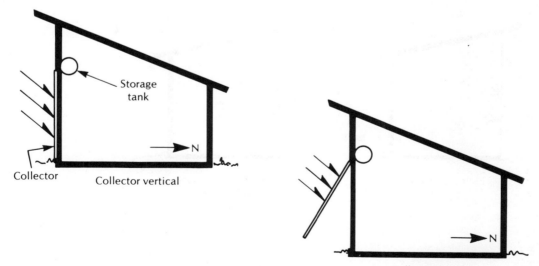

Collector vertical

Collector tilted

Scheme S–46c

One could provide two, instead of one, large tanks. One of these would be above the small compartment and the other below. During the winter only the upper large tank would be used; the other would be disconnected (but left full of water to add to the thermal capacity of the room). During the summer, the lower large tank would be used and the other would be disconnected.

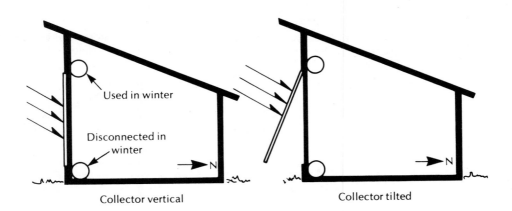

Collector vertical   Collector tilted

Scheme S–46d [of 10/12/78]

Scheme S–46 calls for use of oblique-ended tanks, which would have to be manufactured specially. One could, instead, use ordinary, square-ended tanks and tilt them, and provide small, flat, high-quality reflectors close to the south ends so as to reflect solar radiation toward those ends. The reflectors, which would be near-horizontal, could be indoors, outdoors, or both. Their tilts could be manually adjusted from month to month.

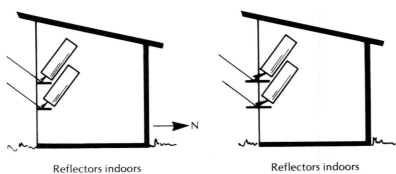

Reflectors indoors   Reflectors indoors
                            and outdoors

Scheme S-46e [of 10/13/78]

As above, but orient the tanks more nearly vertically and use curved reflectors. The tanks then take up less useful space in the room, and the reflectors can collect somewhat more solar radiation. Nighttime heat-loss to the outdoors from lower ends of tanks is almost 100% eliminated. No valves at all are needed, and no internal insulating plates.

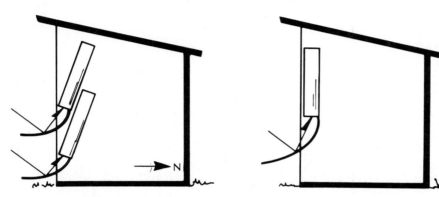

Tanks steeply sloping;   Tanks vertical;
reflectors curved   reflectors curved

PART 2

Combination Passive-and-Active-Systems

INTRODUCTION

It is hard to see why, for any building, a designer who is committed to using an active solar heating system would specify *just* an active system. Nearly always the use of combination passive-and-active seems preferable.

TWO FAMILIES OF COMBINATIONS

There are two kinds, or two families, of combination passive-and-active systems:

A Family that includes just one collection system and

   1. collection is active and distribution is passive, or
   2. collection is passive and distribution is active.

B Family that includes two collection systems, and

   1. the collection systems are optically in parallel, i.e., with independent collection apertures, or
   2. the collection systems share the same collection aperture sequentially: first one system uses it, then the other; that is, there is convertibility, or
   3. the collection systems share the same collection aperture simultaneously: they are optically in series.

Types A1 and A2, called hybrid, are well known. Certainly they have big futures. Type B1 also is well known and has a big future. Types B2 and B3, involving collection-aperture sharing, are scarcely known at all. But they have some important advantages. Will they too have big futures?

In this Part we deal with all of these types. But we deal only with those designs that do not involve use of curved surfaces, i.e., do not concentrate the radiation.

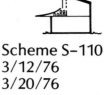

SYSTEM THAT EMPLOYS A SEMI-ABSORBING, SEMI-TRANSPARENT INTERCEPTOR SHEET AND ACCOMPLISHES PASSIVE AND ACTIVE COLLECTION SIMULTANEOUSLY AND OPTICALLY IN SERIES

Scheme S-110
3/12/76
3/20/76

SUMMARY

Figure 1 shows a conventional type of passively solar heated house that performs well but has five noteworthy drawbacks.

Figure 2 shows the proposed scheme, which, by making use of outdoor near-horizontal reflectors, indoor vertical semi-absorbing semi-transparent interceptor sheets, sheet-like airflow, and contiguous storage bins, avoids these drawbacks.

DRAWBACKS TO A CONVENTIONAL, PASSIVELY SOLAR HEATED HOUSE

Figure 1 shows a conventional, two-story, passively solar heated house in Massachusetts. The entire 30 ft.   8 ft. south wall of each story is of glass and is double-glazed. Floors and walls are of thick masonry. The eaves exclude direct radiation in summer.

In 1960 N. B. Saunders built such a house in Weston, Mass. The cost of the passive solar heating system was almost negligible. Nothing of significance has ever gone wrong with the system. In a typical year the system has provided 50% or 60% of the winter's heat need.

But this general type of solar heating system has, or can have, these significant drawbacks:

1. The high-thermal-conductance areas—the south windows—are very large and consequently the heat-loss in midwinter is large. In January as a whole the heat-loss via the windows may exceed the energy received via the windows. Such loss could be greatly reduced by installing internal, insulating shutters on very cloudy days and at night. But 8-ft.-high shutters are hard to manipulate and store.

2. On sunny days in October, November and December, too much radiation enters the rooms. The rooms get too hot and the occupants open windows and doors in order to cool off. This wastes heat.

3. In cold sunless periods the rooms cool off fairly quickly—in 15 hours, say.

4. On sunny days in September through March, too much *light* enters the rooms. The occupants may be oppressed by glare. Also, rugs may fade.

5. The eaves come into play too early in the year. Even in April they cut off nearly all direct radiation. The rooms tend to be too cold then. Yet if the eaves are made to project less, they cut off too *little* direct radiation in August and September.

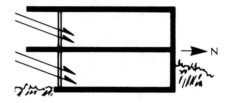

Fig. 1. Conventional passive system.

RATIONALE OF PROPOSED SCHEME

I propose to avoid these drawbacks by using:

- South windows that are only 5 ft. high, with the consequence that the heat-loss area is reduced and the use of shutters is facilitated.

- Crude reflectors that almost double the amount of energy collected per square foot of south window. (Cf. Mathew House in Coos Bay, Oregon, and the analysis by McDaniels et al. in *Solar Energy* of Nov. 1975. See also the article by F. Vignola et al. "Diffuse Radiation Contribution to the Useful Heat Output of a Reflector-Collector System", *Proceedings of the 1978 Annual Meeting* Am. ISES, Vol. 2.1, p. 198.)

- Transparent, gray interceptor sheets (just inside windows) that intercept much of the solar radiation and become hot.

- Storage bins, immediately adjacent to the interceptor sheets, that receive (via sheet-like flow) air that has picked up heat from the interceptors.

- An eaves-and-awning system that blocks intense radiation in hot months only.

PROPOSED SCHEME

Figure 2 shows a cross-section of a 30 ft. × 30 ft., two-story house in which the proposed solar heating system is incorporated. The south wall of the second story consists mainly of a 30-ft.-wide, double-glazed, window that is only 5 ft. high. Adjacent to the lower edge of the window is a 16-ft.-wide, slightly sloping, crude reflector employing aluminum. Four inches north of the window there is a vertical *interceptor* sheet of transparent gray plastic that extends to within 1 ft. of the ceiling. The interceptor absorbs 60% of the incident solar radiation, reflects a little, and transmits about 30%. The visual transmittance is somewhat greater than that of typical sunglasses; thus ample light is admitted to the rooms, the occupants have no shut-in feeling, and they can look out at the scenery easily. As explained below, air near the ceiling (the hottest air in the room proper) is made to flow south and enter the 4-in. space (plenum) between interceptor and window, and flow downward, collecting heat from the south face of the interceptor. Adjacent to the other (north) face of the interceptor there is a slow, gravity-convection, *upward* flow which also collects heat from the interceptor; this flow eventually goes over the top and joins the downward flow in the plenum.

Adjacent to the base of the interceptor, i.e., in a hollow region below the reflector, there is a 30-ft.-long, 8-ft.-wide storage bin which receives the above-mentioned sheet-like flow of hot air from the plenum. The bin is loosely packed with 1500 one-gallon plastic bottles (costing $500 in all) containing 5 tons of water. From here,

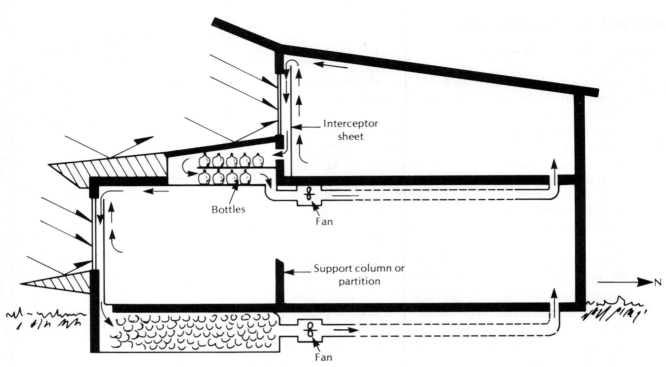

Fig. 2. Proposed system

the air flows north along a channel and enters the room at the north (coolest) side. The flow is maintained by a fan situated in the channel, at the south end; the fan runs whenever the temperature in the plenum exceeds 75°F (unless manually controlled otherwise). Note that practically 100% of the energy intercepted by the interceptor is transported *away from* the room proper *to* the storage bin; on a sunny day the temperature of the bin may exceed that of the room proper by 10 F degrees (a guess).

The *first* story also has a 30-ft.-wide, 5-ft.-high, double-glazed window, and this is served by a 6-ft.-wide sloping reflector. This window, also, is equipped with an interceptor, and warm air from the interceptor is sent to a below-floor storage system—a 25-ton bin of 4-in.-dia. stones. Stones are used, rather than water-filled bottles, for convenience and to save money. The greater weight of the bin of stones is of no consequence here. The thermal capacity of this bin is about the same as that of the other bin. Air from the bin is driven (by a fan) along a channel or duct leading to N edge of the ground floor.

The house itself is very well insulated and has massive masonry floors and walls. The insulation is on the outside of the walls. Floors, and interior walls (not shown), are channeled to facilitate heat input and output. The undersurfaces of the eaves are sloped so as to admit direct solar radiation even in April (but not after mid-May). Additional radiation-excluding means for use in summer are discussed below.

Large portions of the interceptor can be unclipped and stored at times when interception is not needed or when brighter indoor illumination is desired. They may be stored all summer.

PERFORMANCE OF SECOND STORY

The amount of direct radiation entering the second-story window on a sunny midwinter day is about (5 ft.) (30 ft.) (1200 Btu/ft.^{2}) = 180,000 Btu. If the reflectivity of the reflector is 75%, about 100,000 Btu of radiation enters this window via the reflector. Total: about 280,000 Btu. If, during the daytime, 80,000 Btu of energy is lost, the quantity to be taken up by the 2nd story store (5 tons of water) and by the second story walls, ceiling, and floor (weighing, together, 25 tons, say, and having specific heat about 1/5 that of water) is about 200,000 Btu. One finds that this is enough to raise the temperature of water, walls, etc., by 10 F degrees. This rise in temperature is acceptable.

PERFORMANCE OF FIRST STORY

The situation is much the same as for the second story. The energy intake is somewhat less because the reflector is smaller. The energy loss likewise is smaller because, just above, there is a warm room instead of cold outdoor air.

OVERALL PERFORMANCE

The system may supply 75% of the winter's heat need (a guess).

AUXILIARY HEAT

Use small electric heaters situated close to, and downstream from, the above-mentioned fans. Run these heaters only at off-peak hours and only when room temperature is below 68°F. There is no furnace, oil tank, or chimney.

EXCLUSION OF SOLAR RADIATION IN SUMMER

Direct radiation is blocked by the eaves and by awnings. Radiation that is reflected by the big second-story near-horizontal reflector is blocked by a vertical, east-west "fence" consisting of a green canvas strip installed, in summer only, along the east-west centerline of the reflector. After midnight each night the bins are cooled by forced circulation of cool night air and on the following hot days the bins can be used to help keep the rooms cool.

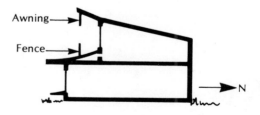

CONCLUDING LIST OF GOOD FEATURES

There are no big inviting areas of glass for vandals to smash. No big sloping glass areas to collect snow, be hit by hailstones, and become too hot in summer. No flowing liquids, pipes, soldered connections, corrosion problems, antifreeze, heat exchangers. No moving parts (except two fans). Virtually no maintenance. Collection efficiency near 75%. Zero collection-start-up time. Performance fully predictable. Principles fully understandable. Windows can be inspected and serviced without need for ladders (you can walk on the reflectors). System works fairly well even if electric supply fails. System is cheap. No furnace needed. Almost no on-peak electric power needed. Excellent performance in summer.

SOME ALTERNATIVES

Have just one storage bin: one extra-large bin below the ground floor. Preheat the domestic hot water by means of a coil within a bin. Use an interceptor that absorbs all of the infrared component, but *little* of the visible component. For auxiliary heat, use a wood stove. Replace the 5 tons of water with 2 tons of phase-change material having a 75°F phase-change temperature, thus roughly doubling the thermal capacity and reducing the volume to one quarter.

SCHEME, USING FIXED COMPONENTS ONLY,
WHEREBY VERTICAL SOUTH WALL OF HOUSE
CAN SIMULTANEOUSLY PERFORM THREE
FUNCTIONS: ACTIVELY COLLECTING SOLAR
ENERGY, ADMITTING DAYLIGHT TO ROOMS,
AND PROVIDING VIEW

Scheme C–70
9/16/74

SUMMARY

Either (1) leave some transparent gaps in the vertical south–wall
collector array or (2) design the collector panels themselves so that
they will be slightly transparent. An optical designer would call
these approaches *parallel* and *series*, respectively.

INTRODUCTION

To maximize the solar–radiation–collection area of a solar–heated
house, one should cover not only the south roof slope but also the
entire south vertical wall with collector panels. See sketch.

But this wrecks the normal functioning of the south vertical
wall. No daylight can pass through it into the room. People in the
room cannot enjoy a view to the south.

Why bother to maximize the collection area? Why try to
achieve near–100% solar heating, rather than, say, 60%? In order
to make it unnecessary to have a furnace room, a furnace, a chimney, an oil tank and to eliminate smell of oil, fire risk, and depen-
dence on regular deliveries of oil. Necessary backup could be provided by a small electric heater or a wood–burning stove.

What can the designer do to resolve this conflict—conflict
between desire to maximize collection area and desire to retain
some transparency of vertical south wall?

There are several ways to resolve the conflict. Some pertinent
schemes are static and some are dynamic. In this essay I deal just
with static schemes.

Collector
panels

PROPOSED SCHEME C–70

Cover most of the south vertical wall with collector panels, but
leave some transparent areas (gaps). An important amount of solar
radiation will enter the rooms via the gaps, providing illumination
and some heating; also, people can gaze out through the gaps.

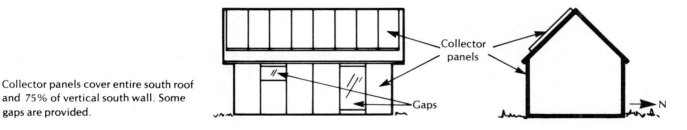

Collector panels cover entire south roof
and 75% of vertical south wall. Some
gaps are provided.

PROPOSED SCHEME C-70a

Cover the entire vertical south wall with collector panels but employ panels that are slightly transparent. Instead of using panels that have 95% absorptivity and 0% transmittance, employ panels that have 75% absorptivity and about 20% transmittance. The amount of energy collected by the panels will be about ¾ as great as with conventional panels. About 20% of the radiation will enter the rooms and will provide some heat and some illumination; people in the rooms can view the scenes to the south comfortably. Note: when you wear 20%-transmittance sunglasses the illumination is still so bright that you may forget you are wearing sunglasses.

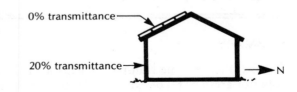

Ideally, the spectral transmittance of the panels would be arranged so that the transmittance is fairly high just in the part of the spectrum where the eye is most sensitive (greenish-yellow portion of spectrum) and is zero elsewhere in the spectrum. A "visual transmittance" of about 30% could be achieved even if the "total solar radiation transmittance" were only about 10%. That is, the panels could collect almost the usual amount of energy, yet they would appear to the eye to be about 30% transmitting.

How would one actually make a partially transmitting collector panel? Use, as "black absorbing surface," not a black-coated sheet of metal, but a gray (or dark green) sheet of transparent glass or plastic. Energy absorbed by the sheet could be picked up by an airstream (at front, or front and back). Conceivably a flow of water could be used, although this would probably greatly interfere with the occupants view. Use of *air* flow is probably best.

The insulating backing of the collector should consist of several spaced sheets of transparent glass or plastic. High-R-value insulation is not needed inasmuch as the heat that leaks through the backing leaks *into* the rooms and helps heat them.

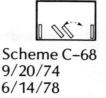

COLLECTION SYSTEM, EMPLOYING HINGED INDOOR PANELS, THAT IS INSTANTLY CONVERTIBLE FROM ACTIVE TO PASSIVE, RETAINING MANY OF THE ADVANTAGES OF EACH

Scheme C–68
9/20/74
6/14/78

SUMMARY

After listing the advantages and disadvantages of employing indoor collector panels that are in fixed positions, I describe schemes that employ hinged indoor panels. By swinging these through a large angle the house occupant can convert the collection system from active to passive. If the axes of rotation consist of 4-in.-dia. vertical pipes, the pipes themselves can serve as ducts for incoming and outgoing air, which reduces complexity and cost. By leaving the panels swung only partially open, the occupant may enjoy many of the advantages of active and passive systems simultaneously.

INTRODUCTION

Advantages of Employing Collector Panels that are Indoors and are Fixed

No antifreeze is needed: the collector never becomes colder than about 60°F.

Because no antifreeze is used, no heat exchanger is needed.

A relatively small amount of insulation on the back of the collector suffices.

Heat that leaks from the collector leaks into the room, helping to warm it.

Collection efficiency is high (because the surrounding air is warm); accordingly, collection at fairly high temperature is feasible; this in turn means that even a relatively small storage system can store much energy.

The collector is not exposed to ultraviolet radiation. Such radiation is blocked by the windows. Accordingly, cheaper panel glazing may be used.

The collector is not exposed to rain or snow. A simpler housing suffices.

Being lightweight and not exposed to wind, a simpler support system suffices.

The collector receives relatively little solar radiation in summer because it is indoors and vertical; accordingly, no components of the collector will ever become very hot and there is no need to use materials that can withstand very high temperatures. Cheaper materials may be used.

Hailstones and vandals cannot reach the collector.

Tax assessors may not notice the collector.

Each panel may be so light that one man can easily lift and transport it.

Because the collector is handy and accessible and is in a warm room, making repairs and adjustments is easy.

The collector largely solves the problem of excessive glare in a room that has very large south windows.

Likewise it largely solves the problem of overheating of such rooms on sunny days.

If suitably designed, the collector may be used at night as a thermal shutter.

In summer the panels could be provided with white faces and could be pressed close against the big south windows to exclude solar radiation and help keep the rooms cool. (Alternatively, the panels could be removed entirely in summer and stored in the basement.)

Disadvantages

The collector panels may block the occupant's view of outdoors.

The collector may block incoming daylight. The rooms may be dark.

The collector may prevent solar radiation from penetrating deep into the room and warming its wall and floor promptly; thus on cold mornings the rooms may remain cold for a long time.

The collector may take up valuable space.

The collector may be unattractive.

PROPOSED SCHEME

Here I propose a scheme that retains most of the above-listed advantages but avoids most of the disadvantages. I propose that the collector panels be mounted on axles so that they can be (a) swung close to the windows, and parallel to them, or (b) turned so as to be more nearly normal to the windows. Intermediate orientations may be preferred under certain circumstances. Of course, some panels could be fixed, parallel to the windows; also, some portions of the south wall could be left permanently free of panels so as to provide light and view.

Whenever the room is cold, during the daytime, the movable panels may be swung so as to admit radiation deep into the room, to warm it promptly. This arrangement may be preferred early in the morning on sunny days and all day on cloudy days.

Whenever, during a sunny day, the room is warm enough, the panels may be swung so as to intercept much solar energy and deliver heat to the storage system.

At night, the panels may be oriented so as to act like thermal shutters.

Note: If the occupants are lazy, forgetful, or are away from home for a few days, no great harm comes from leaving the panels fixed. Nothing is damaged.

Often, during the day, the occupants may prefer to turn the panels to an intermediate orientation, such that much radiation is absorbed, yet illumination and view are retained. In the morning an orientation of 45 deg. may be optimum, and in the afternoon 135 deg. may be optimum.

Instead of swinging about an axle, a panel could slide laterally (like a patio door), slide upward (like a garage door), fold up (like a folding shutter), compress (like an accordion), or unroll (like a window shade). Here we deal only with schemes employing hinges or axles.

SPECIFIC DESIGN

The panels are of air type. Each is 7 ft. × 3 ft. × 5 in., single–glazed with cheap plastic, and includes a black absorber sheet. Much room is left for circulation of air within the panels, and baffles are included to make the airflow here turbulent. On the north face of the panel there is 1 inch of insulation. At top and bottom, near one edge, there are axles which consist of 4–in.–dia. pipes. One serves as inlet duct and the other as outlet duct. Each panel may be swung through a large angle, such as 45, 90 or 135 degrees, manually. A suitable storage system (bin–of–stones in the basement, say) is provided, and a suitable blower, controls, etc.

The sketches show several of the panels in open and closed positions.

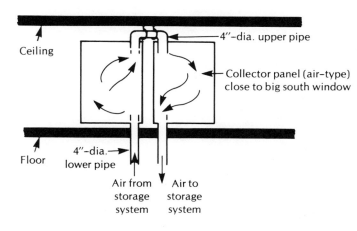

Fig. 1. Vertical cross section (looking south) showing two vertical air-type panels close to the south window. The storage system is in the basement. The two panels are pneumatically in series. The inlet and outlet pipes serve as axles also.

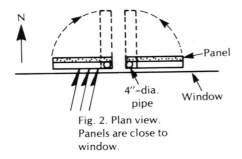

Fig. 2. Plan view. Panels are close to window.

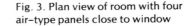

If the axles of the two eastmost panels are along the east edges of the panels and the axles of the two westmost panels are along the west edges, and the panels are oriented optimally, especially good performance is achieved in the early morning hours and late afternoon hours. Practically all of the direct radiation transmitted by the big south window strikes the panels, yet much diffuse radiation reaches the center of the room.

Fig. 3. Plan view of room with four air–type panels close to window

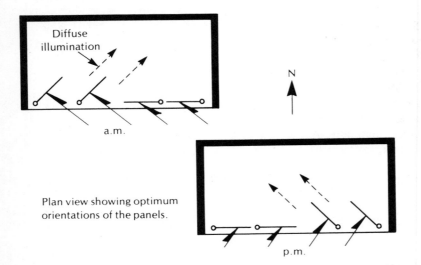

Plan view showing optimum orientations of the panels.

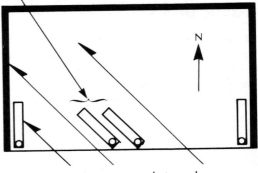

Fig. 4. Same, except that panels have been turned to admit direct solar radiation in morning

OTHER COMMENTS

The north side of each panel could be painted white to make the room look lighter and more cheerful.

When a panel is swung against an end wall the entire panel could be concealed by a white curtain that is rolled down from a fixed–position roller situated just above and attached to end wall or ceiling.

Provision could be made for insuring, when (at night) the panel is pressed against the window to act as a thermal shutter, that no room air circulate between panel and window.

Swinging the panels toward or away from the windows could, of course, be automated, with the aid of timers and/or solar radiation sensors and temperature sensors. The added cost and complexity would not be justified, ordinarily.

If the panels are of water-type, the axles (doubling as water inlet pipe and water outlet pipe) might be only 1 inch in diameter. Probably no antifreeze would be needed.

Some of the panels might be constructed so as to be translucent, rather than opaque.

The overall solar heating system of the given house might include, in addition to the above-discussed hinged panels, some fixed panels on the roof and perhaps some fixed passive solar heating also.

Note on Trial Construction and Use

I learned on 9/14/78 that C. G. Wing of Cornerstones has, in the last few years, actually built a system much like the one described here. The air-type collector panel could be swung about vertical axles that consisted of 4-in.-dia. pipes carrying input and output air. The system is said to have performed well.

MODIFICATIONS

Scheme C-68a

Here the axles run along the vertical centerlines of the panels, instead of along the panel edges, and the axles are moved about 1 ft. north, so as to be about 1 ft. from the vertical south window. These panels can be turned slightly clockwise or slightly counterclockwise. If, during the morning, they are turned slightly counterclockwise, they intercept practically all of the direct solar radiation transmitted by the big vertical south window and yet allow much diffuse radiation to penetrate deep into the room and allow a fair amount of view. In the afternoon they may be turned slightly clockwise to provide correspondingly helpful performance.

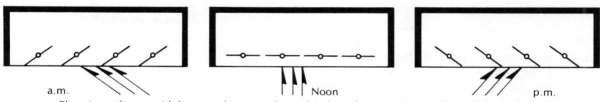

Plan views of room with four panels mounted on axles along the vertical centerlines. The panels are at the orientations that maximize the amount of radiation received by the panels during the indicated part of the day.

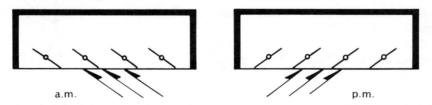

Here the panels are oriented to *minimize* the amount of radiation received by the panels and *maximize* the amount of radiation that penetrates deep into room.

Scheme C–68b

Here the axles are horizontal and the collectors are arranged somewhat like a giant venetian blind. If each panel is turned upward against the window, it receives much solar radiation and, in addition, acts at night as a thermal shutter. If each is lowered (turned) about 45 degrees it receives much solar radiation and in addition allows much diffuse radiation to penetrate deep into the room, and the occupants enjoy some view of the outdoors. If the panels are lowered much farther, nearly all of the direct and diffuse radiation penetrates deep into the room, warming it promptly.

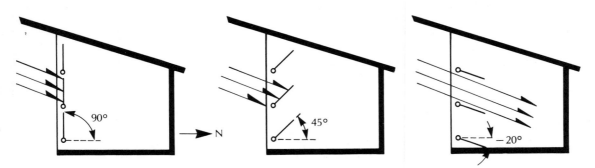

Diagrams showing cross sections of house, looking west, with the panels tilted 90°, 45°, and −20° from the horizontal.

Scheme Proposed by N. T. Pierce

In about 1974 N. T. Pierce of the Massachusetts Institute of Technology invented a collection system consisting of a set of vanes that are curved, to concentrate the direct solar radiation, and include black tubes to absorb the radiation. As indicated in the accompanying sketches, each vane reflects radiation toward a black tube recessed in the back of the next higher vane. The tubes, which have an east–west direction, serve as axles for the vanes. Because concentration is used, energy is collected at especially high temperature. Diffuse radiation enters the room via the spaces between the vanes, to illuminate the room. Heat escaping from the black tubes helps warm the room. At night the vanes can be turned so as to constitute a thermal shutter. Being indoors, the vanes are not exposed to wind or rain; accordingly they can be of lightweight, inexpensive construction. No tracking is needed (other than a slight readjustment every week or two) for efficient collection throughout a 5–hour midday period; with tracking, the period of efficient collection may be as long as 7 hours. Pierce's system, on which a US patent is about to be issued, is described in *Solar Energy, 19,* 395 (1977), *Solar Age,* Feb. 1978, p. 18, and in *Popular Science,* Nov. 1978, p. 19.

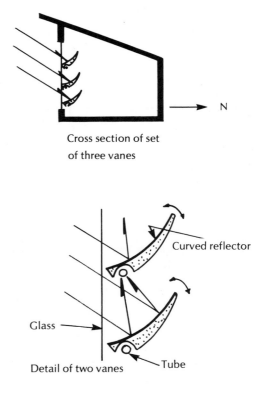

Cross section of set of three vanes

Curved reflector

Glass

Detail of two vanes

Tube

COLLECTION SCHEME, EMPLOYING LIGHTWEIGHT BLACK ABSORBER SHEETS THAT CAN BE LOWERED OR HOISTED, THAT IS INSTANTLY CONVERTIBLE FROM ACTIVE TO PASSIVE

Scheme C–70c

PROPOSED SCHEME

The vertical south wall of the house consists mainly of vertical air–type collector panels the hot air from which is circulated to a storage system in the basement. The north faces of the panels form the south wall of the room. The lower half of each north face of a panel consists of an opaque insulating sheet, but the upper half consists of two sheets of ordinary glass. That is, the upper half, besides providing much insulation, is transparent. Situated centrally between the north and south faces of each panel there is a sheet of black corrugated aluminum. The sheet in the upper half of the panel is separate from that in the lower half. The lower sheet is fixed whereas the upper one can be lowered 3 ft., or hoisted, by means of a rope and pulley system. When it is lowered, solar radiation can penetrate deep into the room; that is, the solar radiation encounters the south–face glazing and also the two sheets of glass of the north face—but nothing else. Solar radiation enters the room and illuminates and warms it. Also the occupants can enjoy a view of the outdoors.

At most times on sunny days both of the aluminum panels are in place (one above, one below) absorbing solar radiation, and a forced flow of air past the front and back of each sheet carries heat to the storage system. Raising and lowering the upper aluminum sheets has no effect on the airflow; the circulation of air to the storage system is in no way changed and the air in the panels cannot escape into the room. The only effect of hoisting an aluminum sheet is that the pertinent big beam of solar radiation heats this sheet (and more energy flows to the storage system) instead of entering the room and heating it.

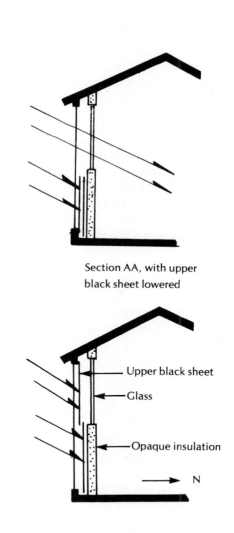

Section AA, with upper black sheet lowered

Upper black sheet

Glass

Opaque insulation

N

Section AA, with both black sheets in place

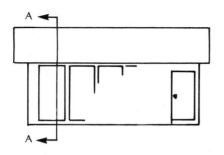

General view of house, looking north

The pulley and rope, and likewise the ducts carrying air to and from the collector, are not shown in the sketches. Guiding springs or ways may be needed to keep the movable black sheet from fluttering in the stream of forced air. Little strength is needed to raise the movable aluminum sheet if a counterweight is provided. The designer could arrange the controls so that one single operation would move the upper sheets in *all* of the panels.

MODIFICATIONS

To remove a large black sheet from its normal position within the panel, slide the sheet west—behind some other black sheet. That is, use a sliding motion rather than a hoisting motion. Sliding could be facilitated by small rollers at the bottom of the movable sheet or by means of suspension cords.

Instead of a rigid black sheet, use a roll-up (or roll-down) black sheet, like an ordinary window shade.

Use a set of rotatable black vanes, shaped and controlled like the vanes of a venetian blind.

Arrange for the entire area of the panel to be made transparent. Use a full-height absorber that can be slid out of the way, or rolled up—again without hurting the pneumatic integrity of the space in which airflow occurs.

Instead of using glass as north face of upper part of collector, use four sheets of cheap, translucent plastic, spaced 3/4 in. from one another. This reduces heat loss at night, but the occupants can no longer enjoy a clear view of outdoors.

Replace the upper black aluminum sheet with a dark-gray sheet of transparent plastic. Then the occupants can at all times obtain at least a dim view of the outdoors. Or, if one insists on using a sheet of black aluminum, cut a 1 ft. × 1 ft. peephole in it.

Design the control system so that all of the movable black aluminum sheets rise into place automatically whenever room temperature rises to 70°F. A thermostat would merely release a catch that allows a previously manually energized spring to come into play. A fully automatic system could be provided, but would cost more.

COMBINATION SYSTEM, EMPLOYING LARGE VERTICAL SOUTH WINDOW AND SMALL INDOOR SPECIALLY-TILTED ACTIVE COLLECTOR, THAT COLLECTS WITH OVERALL EFFICIENCY OF 90%, DELIVERING ABOUT TWO THIRDS OF THE ENERGY PASSIVELY AND ABOUT ONE THIRD ACTIVELY TO STORAGE SYSTEM AT 200°F

Scheme C–69
7/26/78

PROPOSED SCHEME

Start with a large living room that has a large, vertical, south-facing, single-glazed window. Install, fairly close to the window (and within the room) an active-type (air or water) collector panel that is double-glazed and has a thick insulating backing. Tilt the panel so that the direct radiation that is reflected by the panel's glazing travels upward and strikes the ceiling, warming it. Notice that, in upshot, 90% of all of the solar radiation that strikes the window enters the room and contributes either to (a) heating the room immediately, passively, or (b) heating the coolant within the panel very hot indeed—to 200°F, say. Note that all of the radiation that is reflected by the panel's glazing is "saved": it warms the ceiling and thus contributes to heating the room. Likewise all of the energy that leaks from the panel by conduction is saved and contributes to heating the room. I assume that the tilted collector is large enough so that, during sunny midday hours, about 30% of the solar energy entering the room is delivered to the storage tank.

If the homeowner were to remove the panel entirely, the same amount of energy would be collected, i.e., about 90% of the energy incident on the big window. But the room would soon become much too hot and the occupant would open windows in order to cool off; also, little energy would be stored for the coming cold night and the occupant would then have to turn on the furnace. If the proposed scheme is used the room will *not* become too hot, the occupant will *not* open windows, and a large amount of energy will

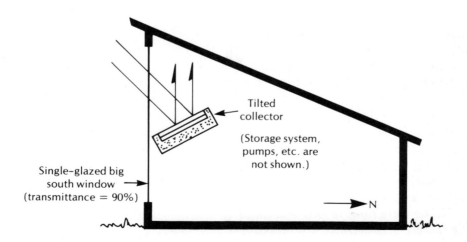

Tilted collector

(Storage system, pumps, etc. are not shown.)

Single-glazed big south window (transmittance = 90%)

N

be stored for future use. Because, at the end of a sunny day, the storage temperature is very high—200°F, say—even a small storage system can store much energy. Collecting at a temperature of 200°F is feasible because the collector is in a warm room and is double-glazed. Double glazing the collector is permissible because it entails no reflection loss: all of the radiation reflected remains in the room, helping warm it.

Note that the collection efficiency of 90% holds even on cloudy days and even on the coldest days. It holds always, instantly: no warm-up period is required.

Thermal shutters or shades could be used at night to reduce heat-loss through the big south (single-glazed) windows. (See also Scheme C-69c described below.)

The essence of the proposed scheme is that,

there are two collectors, namely the room and the panel;

one is much smaller than the other and is inside it;

the smaller one is mounted in such a place, and with such a tilt, that it is three-way-thermally-isolated from the outdoors (no energy from the little collector can reach the outdoors; none of the three energy-transfer mechanisms—conduction, convection, radiation—carry energy to the outdoors);

people live in the larger collector (the room), and accordingly all of the energy this collector gains from the smaller collector helps keep the people warm;

the two collectors perform very different functions: the larger one stores energy at low (70°F) temperature, contains people, and contains the smaller collector; the smaller collector helps store energy at very high temperature;

there is 100% salvage of energy escaping from the little collector.

MODIFICATIONS

Scheme C-69a

Instead of using one panel, use many, arranged one above another, all tilted so that all of the reflected radiation strikes absorbing surfaces within the room and thus contributes to heating it. Perhaps as much as 40% of the solar energy that enters the room will end up in the 200°F storage system.

Scheme C-69b

As above, except employ, on the panels, *triple* glazing, and deliver energy to a storage system at 250°F—hot enough to perform a wide variety of functions, such as running an air cooling system or a heat engine. Note that, in a sense, there is *no penalty* associated with using 2, 3, or more glazing sheets on each panel. The reflection loss by the panel increases; but all of the reflected energy contributes to warming the room. That is, the overall collection efficiency is *still* 90%, thanks to use of little–collector–within–large–collector and use of 100% salvaging of heat leaking from the little collector.

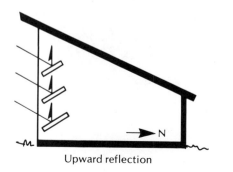

Upward reflection

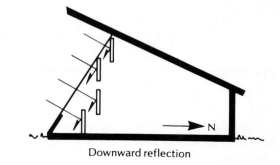

Downward reflection

Scheme C–69c

As above, except mount the panels on horizontal axles that are close to the vertical window. At night, rotate the panels until they lie flat against the window, insulating it. Auxiliary insulating panels, parked during the day on the underside of the tilted panel, could be deployed at night to complete the insulating of the window.

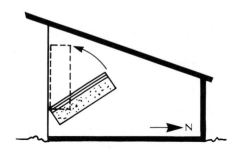

Panel that can be
pressed flat against
window at night

THREE-FUNCTION MULTI-LAYER QUILT ASSEMBLY THAT CAN BE [1] ROLLED UP TO ALLOW PASSIVE SOLAR HEATING OF THE ROOM, [2] ROLLED DOWN AND EXPANDED TO CONSTITUTE AN INDOOR AIR-TYPE COLLECTOR, OR [3] PRESSED AGAINST THE ADJACENT WINDOW AT NIGHT TO INSULATE IT

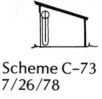

Scheme C-73
7/26/78

PROPOSED SCHEME

The equipment includes a 6-in.-diameter hollow roller that serves not only as a roller but also as collector inlet duct and outlet duct. Also it includes a quilt assembly: a 2-in.-thick, fluffy, compressible quilt proper and a transparent plastic sheet and a black plastic sheet; the assembly is bag-like, and communicates with the interior of the roller, i.e., with the ducts. The storage system, of conventional type, is situated in the basement.

At the start of a cold sunny day, or on a cloudy day, the assembly is rolled up, i.e., out of the way. Solar radiation penetrates deep into the room and warms it.

Later on, on a sunny day (when the rooms are already hot enough), the assembly is unrolled, and the blower is started, expanding the assembly which now performs as an indoor air-type collector. The solar radiation that passes through the transparent sheet strikes the black plastic sheet and is absorbed by it. The quilt proper constitutes the insulating backing of this collector. Much heat is transported to the storage system by the forced flow of air.

At the end of the day the roller is rotated ⅓ revolution so as to bring the top of the quilt proper very close to the vertical south window; thus the entire quilt hangs close to that window and insulates it. The quilt edges are sealed by any suitable means, such as one of the means described in my book on thermal shutters and shades.

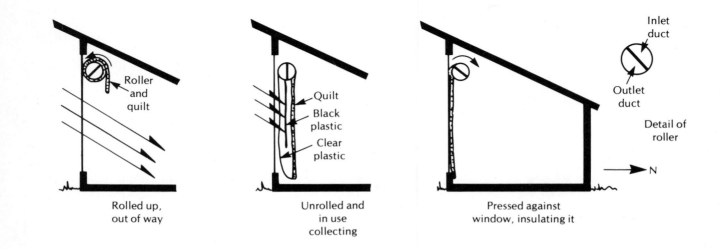

Rolled up,
out of way

Unrolled and
in use
collecting

Pressed against
window, insulating it

Inlet
duct

Outlet
duct

Detail of
roller

N

DISCUSSION

This scheme makes multiple use of the major components. It also provides an indoor collector that can be almost instantly deployed or instantly removed (rolled up). Such collector is very useful: it prevents the room from becoming too hot on sunny winter days, it sends much heat to storage, it collects efficiently (because it is in a warm room), it can be of extremely cheap construction (because it is protected at all times, i.e., is indoors), it can be flexible and thus can be rolled up to allow solar radiation to penetrate deep into the room to warm it promptly. The overall collection efficiency is very high (about 80%) because practically all of the energy that escapes from the collector remains in the room; the efficiency remains high even on cloudy days and on very cold days. The quilt serves the important added function of insulating the window at night.

MODIFICATIONS

Scheme C–69a

Instead of getting the collector out of the way by rolling it up (which requires that the roller turn through several revolutions and entails complications as to bringing air in and out without leakage), merely *fold it up*. One could attach one or more horizontal battens to the quilt assembly, attach ropes to the battens, and using pulleys, hoist up on the battens to bring a large portion of the quilt assembly close to the ceiling. See sketches. The "roller" now serves merely as support and as ducts; it no longer rolls. Two smaller-diameter, side-by-side ducts could be used. At night the quilt assembly could still be held snugly against the window.

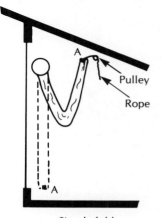

Simple fold

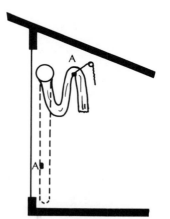

Less simple fold

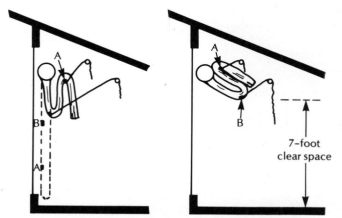

Two-step process: haul up on A, then B

Scheme C–69b

Here the design is, in a sense, inverted. When the quilt assembly is not in use, it may be lowered—collapsed onto the floor, or into a recess in the floor (suggested by Alice Shurcliff). When the house occupant wishes to collect solar energy and deliver it to the storage system, he hoists up the free portion of the quilt assembly (by means of battens, ropes, pulleys). The base of the assembly is fixed and is connected to ducts that pass through the floor into the basement. Just as a sailor hoists his sail when he wants to travel, the solar-house occupant hoists his collector when he wants to store energy.

There is a tendency for especially hot stagnant air to accumulate in the upper part of the deployed collector. But this can be overcome by using turbulent flow of air or by employing a flexible, collapsible upward-extending duct within the collector.

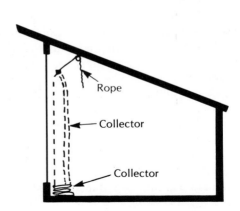

Rope

Collector

Collector

PENTHOUSE SYSTEM EMPLOYING STACK OF WATER–FILLED CONTAINERS, A LARGE SINGLE–GLAZED WINDOW, AND A COLLAPSIBLE 4–IN.–THICK THERMAL CURTAIN THAT AT NIGHT IS HOISTED INTO POSITION TO INSULATE THE WINDOW

S–139
9/30/76
8/30/78

PROPOSED SCHEME

Near the north edge of the roof, which slopes 6 deg. upward to the north and is covered with shiny aluminum, there is a 36–ft.–long, 6–ft.–high penthouse the major axis of which is east–west. Most of the space in the penthouse is taken up by water–filled containers: 1000 one–gal. bottles and 20 55–gal. drums, the former having large aggregate surface area and providing quick uptake (or quick output) of heat, and the latter having great mass and providing much long–term storage. The penthouse is heavily insulated on all faces except the south, which is single–glazed.

On sunny days much direct radiation, and much reflected radiation, passes through the window and strikes the containers and is absorbed by them. At the end of each day, an insulating curtain is hoisted into place immediately north of the window, to insulate it. The curtain includes four layers, each ¾ in. from the next and thus has an R–value of about 10. The four layers are attached to a stiff bar at the top; this may be hoisted or lowered by means of ropes and pulleys.

When the rooms need heat, room air is circulated through the penthouse (to pick up heat from the bottles and drums) via small, inexpensive, flexible ducts that have little or no insulation. A low–power blower is used.

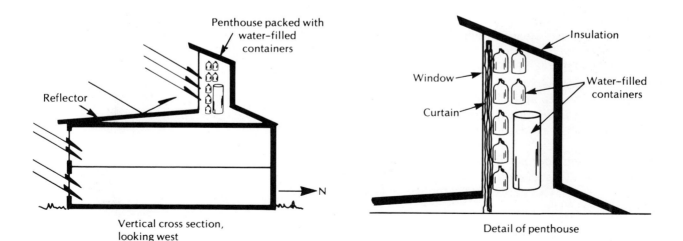

Vertical cross section, looking west

Detail of penthouse

98

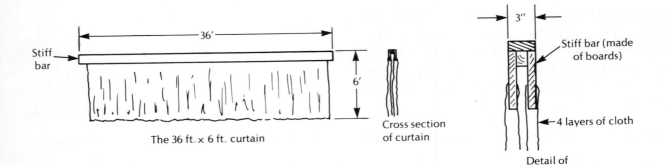

Stiff bar

36'

6'

The 36 ft. × 6 ft. curtain

Cross section
of curtain

3"

Stiff bar (made
of boards)

4 layers of cloth

Detail of
cross section
of uppermost
part of curtain

COMMENTS

Collection efficiency is very high, mainly because a large-area
reflector is used. The reflector greatly increases the solar energy
input to the penthouse, but adds *nothing* to the losses. (See article
by McDaniels et al. in *Solar Energy*, Nov., 1975.) Another reason
for the high efficiency is that the penthouse window is single glazed;
single glazing is permissible because a high-R insulating curtain
covers the window at night or on heavily overcast days.

Notice that the system has *zero* collection-start-up time. As
soon as the solar radiation reaches the penthouse, it is absorbed
directly by the water-filled containers, heating them. Thus collection proceeds with high efficiency even if the solar radiation is
interrupted several times an hour by thick clouds.

The ducts, being entirely within a warm room or the even-
warmer penthouse, need little or no insulation. Slender ducts, and a
low-power blower, suffice because they are used only for heat output, and heat-output may continue throughout 12 or 24 hours per
day, with the consequence that a low rate of heat-transfer is adequate. (When ducts are employed in assisting solar energy *input* to
a storage system, order-of-magnitude-greater energy transfer rate
is needed because intense solar energy input occurs only about 4
hours out of every 24, on the average. A key feature of the present
scheme is that no ducts are involved in energy input to the storage
system.)

There is good access to the penthouse from all sides. It can be
built by carpenters with no special training and is easy to service in
summer or winter.

The curtain assembly is of low-technology type and can be
built locally without difficulty.

No antifreeze is needed in the filling of the bottles and drums.
Their thermal capacity is so large, the penhouse is so well insulated,
and so much radiation is received each week that freeze-up will
never occur.

Because the system does not involve the basement (if any) it
may be ideal for retrofit installation on buildings that have approximately horizontal roofs. Ordinarily, added timbers to support the
several tons of water will be needed.

PART 3

Active Systems
[Non-Concentrating]

INTRODUCTION

Here I discuss active-type solar heating systems that do not involve concentrating the radiation. Concentrating systems are discussed in Part 4.

I make little mention of systems using ordinary air-type or water-type collectors because they are already well known and most of them are too complicated and too expensive. Most water-type collectors include large numbers of rigid tubes, connections, edge strips, seals, etc., and these entail a host of dangers and complications. A formidable list of complications is presented in the first of the following essays. H. E. Thomason's system, however, avoids most of the complications. So does the extruded-roll-unroll-and-glue system developed by Bio-Energy Systems, Inc. These systems, and others, are analyzed in later essays.

Most air-type systems require great lengths of large-diameter, heavily insulated, ducts. A proposed scheme using no ducts is described; it employs sheet-like flow only. Many systems employ bins-of-stones, although water-type storage is far more compact. A proposed way of using a water-type storage system with an air-type collector is described.

UNWISE DESIGN DECISION MADE AT THE OUTSET BY MOST MANUFACTURERS OF WATER-TYPE COLLECTORS, AND REASONS FOR BELIEVING H. E. THOMASON'S DECISION WAS A WISE ONE

8/2/78

SUMMARY

In starting to design water-type collectors for use in solar heating systems in cold climates, most designers apparently have made the basic decision not to have a large area of water in contact with air. They decided, that is, that the water in the collector must be confined within pipes.

They reasoned, I suppose, that hot water exposed to air will evaporate rapidly, evaporation produces cooling, and cooling is just what you don't want in a solar heating system.

Was this decision wrong? Was it a serious economic handicap to the manufacturers involved? In many cases, yes. Here I marshal arguments in support of this dire thesis.

Harry E. Thomason made the opposite decision: in designing his trickling-water-type collector, he accepted having a large area of water in contact with air. Has that decision proved sound? I think so, and I give about 15 reasons for thinking so. Properly used, his collectors perform well; they avoid about a dozen complexities and pitfalls; and they cost, installed, only about ½ or ⅓ as much as typical, high-quality, conventional collectors.

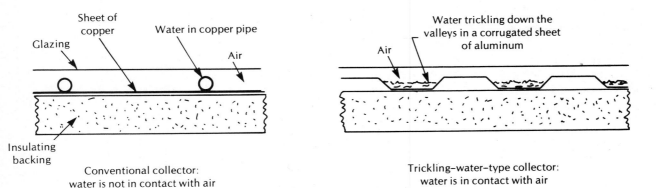

Cross sections of portions of conventional collector and trickling-water-type collector (Not to scale)

DETAILS

Consider an inventor who is starting to design a water-type collector for use in cold regions. Although he knows that use of pipes in the collector can lead to many problems, *he makes the decision to use pipes.*

This decision—to incorporate many rigid pipes in the collector—starts him down a long and agonizing road. Difficulty after difficulty arises. Each is overcome but at a significant cost in complexity and money.

Here is a list of the difficulties. The list is long (unfairly long?) because I have tried, as a stimulating mental exercise, to ferret out *all* of the difficulties: major and minor, direct and indirect.

The piping had better be of copper. Aluminum might corrode in 10 or 15 years.

Extreme overpressure may sometimes occur and, accordingly, the wall thickness of the copper tube must be generous. Also overpressure-averting devices may be needed.

There will be a great many joints, or connection, because, for efficient energy collection, the pipes must be in parallel. There are many pipes, hence many joints.

The joints must be made with great care because high pressure may sometimes occur, certain coolants have great tendencies to leak, the panels are relatively inaccessible, and to make repairs is awkward.

A two-part absorber must be used: fins of aluminum (or very thin copper) and tubes of thick copper. Joining the parts is difficult; close thermal bond is needed; yet differential thermal expansion may pose a threat.

To keep the number of pipes from becoming fantastically large, the designer must place the pipes fairly far apart. Heat within the fin may have to flow as much as 1 or 2 inches in order to reach a pipe. This slightly reduces collection efficiency.

Dirt may clog a pipe during assembly or later and may halt the flow of coolant in this pipe. If a pipe does become clogged, this fact may escape notice for months or years. Meanwhile the performance of the collector suffers. To unclog a pipe may be difficult.

The risk of freeze-up and rupture of pipes is great and must be avoided by one of these procedures:

Use drain down. But this presents problems:
Inlet of air (or nitrogen) must be arranged.
Drain-down could be defeated by air-lock in a pipe.
It could be defeated if a pipe is clogged with dirt.
Filling a pipe with air can sometimes aggravate corrosion problems.
If some defect in the control system or hydraulic system were to prevent drain-down, serious freeze-up might occur.

Use ethylene glycol. But this presents problems:
The material is poisonous. Special precautions must be taken to make sure it never gets into the main water supply system.
It is extremely prone to leak.

It degrades chemically if kept at high temperature for a long period; occasional testing, or complete replacement, is needed:

It has lower specific heat than water.

It is expensive. If, accidentally, it all leaks onto the ground, the owner must buy a new quantity.

If it somehow becomes diluted with much water, freeze-up can occur.

Use silicone oil. But this presents problems:

It has such high viscosity that the designer may feel compelled to order pipes, fittings, and pumps of extra-large size.

Its specific heat is only about half that of water. Thus faster flow rate (and more pumping power) is needed.

Its surface tension is so low that it may leak unless all joints are extremely tight.

Its cost is high—about $2 per pound or $16 per gallon.

Corrosion risk can exist unless silicone oil is used as coolant.

Stagnation temperatures in summer could cause damage. If the coolant is water, the water might boil and cause rupture. If the coolant is ethylene glycol and water, the ethylene glycol may deteriorate chemically; also boiling might occur.

Because the absorber assembly must be made of expensive materials and must be assembled with great care, it had better be made in a high-technology factory. This entails:

Paying enough to cover the manufacturer's profit, the distributor's profit, etc.

Shipping the equipment many hundreds of miles, in a special truck, say.

Crating may be required in some instances.

The panels must be of small size in order that they may be applicable to large buildings or small buildings.

Therefore the weight of the frame relative to the weight of the panel proper is large; and a similar argument applies to the *cost* of the frame—it too is relatively high.

The panel may be too heavy for convenient lifting. Special hoists may be needed, especially when mounting the panel on a high steep roof.

When the panels are installed on such a roof, many connections must be made and must be made very reliably.

Therefore experts may be needed to make the connections, e.g., high-salaried workers whose home base is far away.

The installation expense may amount to a considerable fraction of the cost of the panels themselves.

Because the panels are small in area and there must be many of them, the total linear amount of panel edge, or framing that must be carefully sealed is very great. At 20 ft. per typical 7 ft. × 3-ft. panel, the total amount of panel edge for a typical installation may be 500 ft. Aside from the weight and cost of so much edge or frame, the sheer task of insuring tightness of seal (of glazing, of insulating backing, etc.) along such an amount of edge is a formidable one.

Because the panels are mass produced in a factory, the temptation to use metal framing is irresistible. But metal is expensive. Also it is a thermal conductor and must be insulated.

In some instances the designer feels obliged to use:

pH control equipment

flow meter

surge tank

air separator

check valve

weep holes in collector panels, to avoid pressure build-up from condensate

If the reader thinks these arguments are overdone, let him read:

The Solar Decision Book by Montgomery and Budkin (Dow Corning Corp., 1978, $10). The authors stress repeatedly the need to use highest-quality, high-performance components and the most durable materials; otherwise, various serious troubles may arise within a few years.

Some Steps to Solving Solar System Problems'', by H. Orlowski, *Solar Engineering*, July 1978, p. 31-34. Scores of pitfalls are indicated.

No wonder conventional systems intended to last 20 years are so very expensive. As a corollary, no wonder the public shies away.

Note: Although the glazing of conventional collector panels is impervious to water, and although the edges are well sealed, the glazing performs one less task than it could perform—besides confining warm air and trapping 4-to-40-micron radiation, it could be used to confine water or water vapor. But it is not. A wholly separate system, involving an enormous number of pipes and joints, is used for this purpose.

THOMASON MADE THE OPPOSITE DECISION

About 20 years ago Harry E. Thomason made the opposite decision—he decided that it *is* permissible to have a large area of water in contact with air.

As nearly everyone now knows, Thomason employs a corrugated black sheet of aluminum and arranges for water to trickle down the valleys of this sheet. He relies on a sheet of glass, situated about ½ in. above the trickling water, to trap or confine the high-humidity air that is immediately above the water. At the start of a sunny day, a pump starts the flow of water to the feeder pipe along the upper edge of the collector: water starts trickling, solar radiation warms the black aluminum and the water, some water evaporates, and the confined air soon becomes almost 100% humid, i.e., becomes practically saturated with water vapor. Throughout the rest of the day there is practically no change in the humidity; some water evaporates and an equal amount condenses. Some of it condenses on the glazing, warming it, and the glazing in turn loses energy to the surrounding air. This is one of the overall heat-losses of the system. But if the collector is used properly, i.e., in conjunction with a storage system and heat distribution system designed to perform well even when the storage system is at only about 90 or 100 or 110°F, collection proceeds (on not-very-cold days) with an efficiency of about 35 to 65%, according to my interpretation of a very recent report by J. Taylor Beard "Engineering Analysis and Testing of Water-Trickle Solar Collector," Final Report, Report ORO/4927-78/1, from University of Virginia.

SOME ADVANTAGES OF THIS SYSTEM

The collector proper includes no pipes, except one distribution pipe along the upper edge.

It includes practically no copper.

No high pressure can be built up; nothing has to be sturdy enough to resist high pressure.

No seals are subjected to any appreciable pressure.

No special drain-down procedure is needed. Just turn off the water-pump.

Air-lock cannot occur.

No antifreeze is needed—no ethylene glycol, no silicone oil, no pH control.

No heat within the corrugated aluminum sheet has to flow more than ¾ inch in order to reach the trickling water. Nearly half of the black surface areas is *directly* in contact with this water.

The wetted surface can be inspected at any time. It could be repainted, should this ever be necessary.

If any valley became clogged by dirt, this fact would be readily apparent.

If a valley became clogged, the trickling water could readily detour around it.

A single-part absorber suffices—a single sheet of black corrugated aluminum.

Because sheets are available in lengths up to 36 ft., a single sheet can extend the entire way from peak of roof to gutter at bottom. Sheets are easily cut to length.

Even a very long sheet is so light that one man can lift it.

Use of wood frame members is permissible. Wood is a thermal insulator.

No horizontal frame members are needed, except at the very top and bottom of the entire system.

The collector can be built by persons with only moderate skill, assuming that proper instructions have been received and a licence has been obtained.

The glazing does triple duty—it excludes rain, snow, leaves, dirt, etc.; it confines moist warm air; it confines 4-to-40-micron radiation—that is, it performs *all* of the confinement tasks.

No heat exchanger is needed.

No flow meter, no surge tank, no air separator, no check valve, and no weep holes are needed.

If, by chance, some coolant escapes and is lost, the financial penalty is negligible. Water is practically free.

The amount of initial testing needed is very small.

DISCUSSION

This general analysis is not entirely fair. In some situations, especially where high storage temperature is necessary and outdoor temperature is very low, use of conventional collectors employing hundreds of pipes or other relatively costly collectors may be highly appropriate. Also, some designers of systems employing many pipes have found ways of avoiding some of the difficulties listed above.

Yet is it not true that, once the designer has elected to use scores of rigid pipes, a long parade of technical problems arises? Is it not true that trickling water systems avoid most of these problems? Has not the low cost and long life of such systems been clearly demonstrated?

If so, was not the initial decision by most designers—that large interface areas between water and air must not be permitted—unwise? Has not that decision often jeopardized their chances of producing a system that is reliable, durable, and inexpensive?

COLLECTOR ASSEMBLED FAST ON-SITE FROM EXTRUDED SYNTHETIC RUBBER SUB-ASSEMBLIES: A HARD-TO-DESCRIBE, LOW-COST SYSTEM ALREADY IN PRODUCTION

10/15/78

SUMMARY

Here we describe an intriguing system developed by Bio-Energy Systems, Inc. and called *SolaRoll*. Remarkably easy to assemble on-site and outstandingly low in cost, it makes great use of synthetic rubber extrusions (strips, mats, etc.) that integrate so many functions that relatively little work remains to be done at the building site. The system, already in production, is hard to describe because it is a strange combination of strange devices assembled in a strange way.

To make the discussion simple, I shall assume that the reader wishes to build a single-glazed, water-type collector for use in a moderately cold region. What this company supplies, for this purpose, is a set of most of the parts, or components, needed, including practically all of the difficult or intricate parts. Using these parts, anyone reasonably familiar with tools can assemble such collector on-site, e.g., on the sloping roof of this house. And he will find that nearly all of the hard part of the work has been done for him at the factory—by an extrusion machine.

The two main components are so strange that their names and functions must be clearly stated at the outset:

Rubber framing strip This is an extruded EPDM-rubber strip that constitutes all of the precise and intricate parts of the sides of the collector frame. It serves as channel for glazing edge and also as clamp, sealant, flashing, and attractive and durable surface finish.

Rubber absorber mat This is an extruded EPDM-rubber assembly that constitutes a complete water-cooled absorber, including tubes, fins (web), and non-selective black surface.

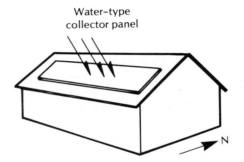

Perspective view of collector panel on roof. Size and shape are here chosen arbitrarily.

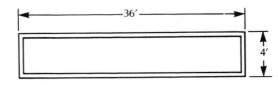

Plan view of panel

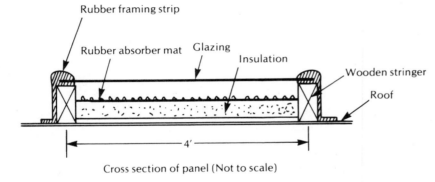

Cross section of panel (Not to scale)

Other components include: header tubes of a novel type that can be connected with the bare hands, fiberglass–and–polyester glazing, and a 5-gal. can of thermosetting adhesive.

The accompanying sketches show how the system might be applied, for example, in the construction of a 36-ft.-long panel.

The complete cost of such a panel, built with the components described here, is said to be of the order of $8/ft.^{2}, assembled, including materials and labor.

INTRODUCTION

I became excited by this set of products, invented and produced by Michael Zinn and Steven Krulick of Bio-Energy Systems, Inc., of Mountaindale Road, Spring Glen, NY 12483, only after being frustrated by the brochures I received: brochures that radiated enthusiasm and promised wonders but left me bewildered as to what, exactly, was being sold. Was it a sealant? A frame? A black tube? A large-area absorber? A heat exchanger? A complete collector? Was it to be used for heating swimming pools, or domestic hot water, or houses?

After writing grumpily for more information, I soon received a detailed letter and a 20-lb. carton of sample components. Later, at a big solar energy exhibition in Boston, I was given a thorough briefing. At last I understood. And I was delighted.

To describe this system is difficult because:

It is so new and strange, involving new devices, new functions, new terminology. The key words are: extrusion, synthetic rubber, roll-up, unroll, clip, and glue.

It consists of a wide variety of components to be used in on-site assembly. It is *not* a complete collector.

It is so versatile that it can be used in many different formats to perform many different functions.

To simplify the story, I choose to explain here how the components are used in just one centrally important application: a single-glazed water-type collector panel. The accompanying sketches show such a panel on the roof of a house. (Panels can be used singly or in groups; they can run up and down the roof or east and west. Here, I use an east-west orientation in order to be able to show how very long a panel can be, if made of the components described below.)

PROPOSED SCHEME AS APPLIED TO A SINGLE-GLAZED, WATER-TYPE COLLECTOR PANEL

To make such a panel one employs the following very special components produced by Bio-Energy Systems, Inc.:

Rubber Framing Strip

Here we consider this strip in detail. The accompanying drawings show its shape, size, and function.

108

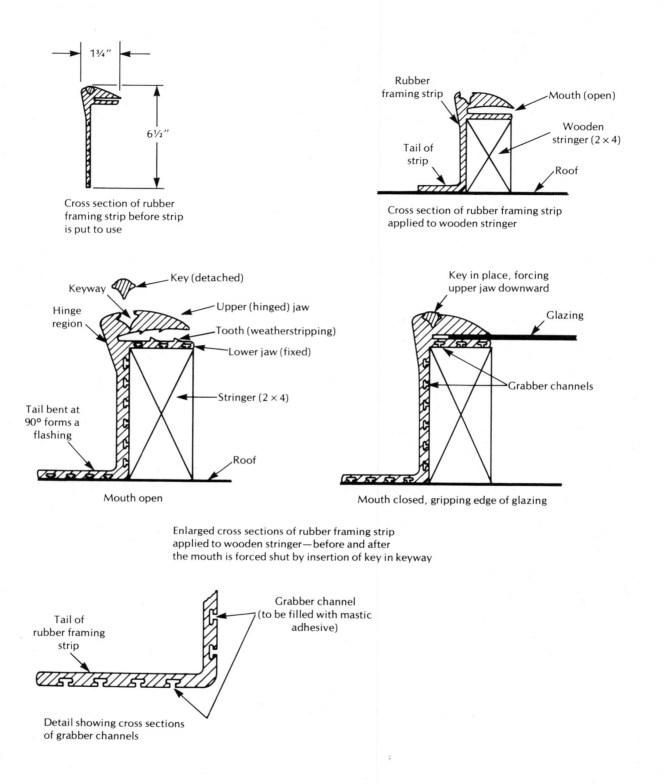

1¾″

6½″

Cross section of rubber
framing strip before strip
is put to use

Rubber
framing strip — Mouth (open)

Wooden
stringer (2 × 4)

Tail of
strip

Roof

Cross section of rubber framing strip
applied to wooden stringer

Key (detached)

Keyway

Hinge
region

Upper (hinged) jaw

Tooth (weatherstripping)

Lower jaw (fixed)

Tail bent at
90° forms a
flashing

Stringer (2 × 4)

Roof

Mouth open

Key in place, forcing
upper jaw downward

Glazing

Grabber channels

Mouth closed, gripping edge of glazing

Enlarged cross sections of rubber framing strip
applied to wooden stringer—before and after
the mouth is forced shut by insertion of key in keyway

Tail of
rubber framing
strip

Grabber channel
(to be filled with mastic
adhesive)

Detail showing cross sections
of grabber channels

The strip, made by extrusion and obtainable in lengths up to several hundred feet, consists of black ethylene propylene diene monomer (EPDM), a material already in widespread use in sunny locations and known to have very long (30 year?) life. It stands temperatures from $-80°F$ to $+375°F$. The strip's "mouth," designed to receive the edge of the glazing, is initially relaxed, i.e., open. After the glazing has been inserted, the mouth is permanently and tightly closed, forming a snug clamp, waterproof edging, and attractive face-plate for the panel frame. To close the "mouth," the builder forces the upper jaw downward and locks it in down position to maintain a strong positive sealing pressure that will endure for decades. How does he force the upper jaw down? With the aid of a long EPDM key which has been shaped, by an extrusion process, to fit in the long keyway in the strip. The key is locked in place by its two tiny lateral ridges, which engage two corresponding grooves in the keyway. Using a special tool (basically a slender steel rod with rounded end) the builder installs the key in seconds. If necessary, it could be removed at any time—with the bare hands—and reinstalled again.

Before the glazing is installed, the rubber framing strips are permanently fastened to the wooden stringers which form the panel frame and have been installed well in advance. The strips are fastened by means of thermosetting mastic—a synthetic, rubber-base adhesive. This is applied (by trowel) both to the strip and to the stringers. The adhesive applied to the strip fills the many tiny grabber channels formed during the extrusion process (two channels per inch). Thus a strong and durable bond is produced. The tail of the strip is flexible and can be bent 90 deg. at almost any location, to suit the width of the 2×2, or 2×4, or 2×6 stringer. The terminal segment of the tail then acts like a flashing to prevent water-leak at the junction of panel and roof. The strip could be attached by nails or screws; but use of adhesive saves much time and avoids producing local stress points.

Rubber Absorber Mat

The entire black central region of a moderate-size collector panel consists of 4.4-in.-wide absorber mats that have been laid out in parallel each having a single 180° turn opposite the header end. The mat is received from the factory in roll form, 600 ft in length. If the panel area is to exceed 220 ft², two or more rolls are used.

The individual 4.4-in.-wide mat, consisting of EPDM and made as an integral unit by extrusion process, includes 6 tubes and intervening thin flat web areas, or fins. Each tube is 3/16-in. ID, 5/16-in. OD, with a 1/16-in. wall thickness. The tubes are 0.7 in. apart on centers. Some of the intervening webs include, on the under side, grabber channels so that the mat can be firmly attached to a flat surface by means of the adhesive mentioned above.

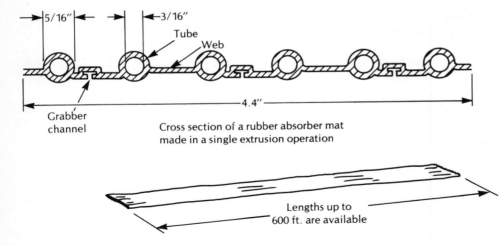

Cross section of a rubber absorber mat
made in a single extrusion operation

Lengths up to
600 ft. are available

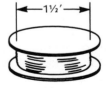

Rolled-up mat

Where the mat is to be turned 180°, the web is stripped away to allow the tubes to turn individually and freely. No adjustment in tube length is needed in view of the cross-over pattern adopted. Of course, 90° turns are made equally easily.

End view of 180°-turn region, showing that the overall thickness, or height, here is less than 1 inch

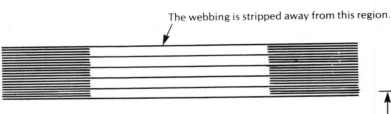

The webbing is stripped away from this region.

Plan-view of mat before a portion is turned through 180°

8.8"

Plan-view of mat after completing the 180° turn. Note cross-over pattern.

The actual layout of the mat, and the arrangement of turns, is such that all of the water inlet and outlet connections are grouped together in a single corner of the collector (say the lower west corner).

Because the tubes are of rubbery material they are not damaged if filled with water and the water is allowed to freeze. Also, there is no corrosion problem. The tubes do not require paint or other special coating—they are "born black"; the (non-selective) carbon black pigment is present throughout the material, not just on the surface. If you pound or scrape the tubes, they still remain entirely black.

Header tube with row
of small holes

Header

Two copper header tubes, each 1 or 1½ in. in ID, are provided, Each has a row of slightly tapered, 3/8-in.-dia. holes spaced to receive the tubes of the mat. Each such tube (by itself) fits loosely through such a hole until an internal rigid tube of Teflon (jamb

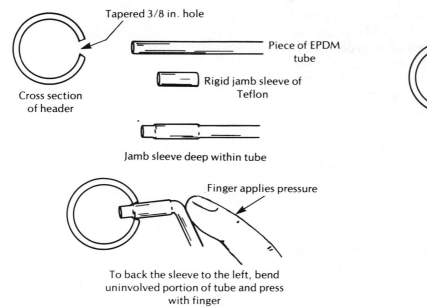

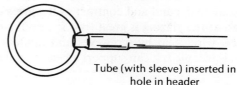

Tube (with sleeve) inserted in hole in header

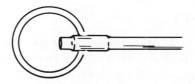

Jamb sleeve has been backed to the left, locking tube in hole in header.

sleeve), ¾-in. long, is forced into place. The above-mentioned taper helps lock the tube-end in place; the joint is watertight and strong. The jamb sleeve is forced far into the tube in advance, then is backed off (by pressure of the finger) into the 3/8-in. hole. The entire locking operation takes only a few seconds per tube. No tools are required.

Ordinarily, successive tubes are used alternately for supply and return, so that the average temperature across the mat is the same at all locations along the mat. Tubes 1, 3, 5, etc. are connected to the supply header and Tubes 2, 4, 6, etc. are connected to the return header. The two headers are situated very close together—within an inch or two of each other—which facilitates installation and makes it easier to insulate the pipes running to the storage system. Arranging to have successive tubes connect to different headers is simplicity itself, thanks to the flexibility of the tubes.

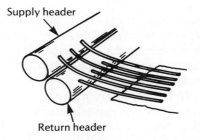

Alternate tubes of mat connected to alternate headers

Coolant

Ordinary water is used, and the headers are drained at the end of the sunny day. The tubes do not drain—if they are horizontal; but no harm is done when and if the water in them freezes. Because no antifreeze or corrosion inhibitor is used, the water is not poisonous. Therefore no heat exchanger is required; the cost and delta-T penalties associated with heat exchangers are avoided.

Glazing

This is a 0.060-in.-thick sheet of fiberglass and polyester with a Tedlar coating. Such sheet, supplied in roll form by any of several manufacturers, is simply unrolled into place; the edges of the sheet are slipped into the open mouths of the rubber framing strips and the mouths are then shut and locked. The mouths provide room for the glazing to expand and contract as it warms up and cools off. If the span of the glazing is great (is 4 ft., say), a criss-cross of stainless-steel wire, held by screweyes and tightened by a single turnbuckle, is installed; it can support a heavy wind or snow load.

Insulation

A plate of rigid, foil-faced insulation is laid down before the rubber absorber mat is installed. The mat is glued to this plate.

EFFICIENCY

I have been told that performance tests made at a well-known eastern test facility show that the overall collection efficiency of a water-type collector such as described above is comparable to that of many collectors of established and familiar type.

Points in favor of the system described here are: (1) no antifreeze is needed, (2) no heat exchanger is needed, (3) thermal capacity is small and accordingly warm-up time is short, (4) the absorptance of the absorber mat holds virtually constant over periods of many years.

A point against the system is that the thermal conductance of the synthetic rubber (EPDM) is far lower than that of copper or aluminum—three orders of magnitude lower. Solar energy delivered directly to a tube must pass through a 1/16-in. wall of EPDM, which presumably imposes a small but real delta-T penalty—of the order of 10 F degrees, according to a rough estimate I have made. Much of the solar energy delivered to the web presumably escapes into the air that is confined beneath the glazing; such energy, serving to warm the air, contributes by giving the tubes a warmer environment and reducing the heat-losses from tubes to air. Persons familiar with the Khanh master-and-slave collection principle will recognize the webs as slaves that help the (tubular) masters.

My expectation is that, overall, the collection efficiency of the collector described here is comparable to that of various familiar collectors and somewhat lower than that of some of the most efficient collectors. However, when one takes into account that this collector needs no antifreeze and no heat exchanger, and takes into account the performance of the solar heating system as a whole, one may expect that the system is closely comparable, in performance, to many other systems. And when one turns ones attention from efficiency (not a legitimate criterion) to Btu's delivered per dollar expended (a highly important criterion), one is likely to rank the present system very high.

SIMPLICITY

It is remarkable how many conventional parts are *not* needed in the present system. In assembling the absorber mats and headers and framing strips:

> no nails or screws (or practically none) are needed,
>
> no tees or elbows are needed,
>
> no battens are needed,
>
> there is no need for soldering or welding, or for sealant or paint.

COST

I understand from the manufacturer that the rubber framing strip costs, at retail, about $2.50 per linear foot. Cost *per square foot of collector panel* is, of course, less if the panel is long and wide. The cost of the rubber absorber mat is about $1 per linear foot. The total retail cost of materials for a 36 ft. × 4-ft. collector is about $5/ft.^{2} Labor costs about $3/ft.^{2} The total cost of a finished collector, installed, may be as low as $8/ft.^{2}, according to information received from the manufacturer.

OTHER COMPONENTS

Certain conventional components of the entire solar heating system are not produced by Bio-Energy Systems, Inc., but must be obtained from other sources.

DISCUSSION

The system seems very promising, especially as most of the intricate and laborious on-site tasks are avoided by intelligent use of extrusion at the factory. Also most of the components can be rolled up for shipment. Durability and appearance seem excellent. Materials cost is low, and labor cost is extremely low.

SOME OPTIONS

The number of options, or variations in design and application, is enormous. I list only a few here.

Panels can be of practically any size and shape, and great economy results when the individual panels are very long (such as 16 ft., 36 ft., or 100 ft.) Panels can be constructed side by side, and, for the intervening wooden stringers, special framing strips with two mouths, facing in opposite directions, are provided. Alternative adhesives—some cheaper, some faster setting—are available, and some of these can be applied with a brush or by sprayer.

Other kinds of glazing can be used: glass, for example. Also, double glazing can be used; installation is especially simple if Rohm and Haas Tuffak is used. Glazing assemblies up to 5/8 in. thick can be accommodated by the extra-large mouth of a new framing strip now being developed.

Larger-diameter, lower-cost headers of polybutylene may be used where there is no danger that very high temperatures will be reached, for example during stagnation in summer. Such headers are used routinely in swimming-pool applications, where there is no need to have any glazing above the absorber mat.

In some applications it is feasible to mount the headers in a warm indoor location—not on the roof. Being indoors, the headers will never freeze; therefore, there is no need for draindown or anti-freeze. The rubber absorber mat is of extra length and extends not only throughout the area of the collector panel but also through a small hole or slot leading to a warm indoor region; the mats may be greatly compressed, or compacted, laterally for this purpose. In summary, we have here the capability of making an entire collector that can be filled with plain water and left filled throughout the winter on a cold windy roof in Maine.

An even more extreme scheme would be to have the EPDM tubes extend all the way into the basement and join the headers there. Chance of freeze-up there is very small. Even more extreme: have the outlet tubes extend all the way into the storage tank; then, for these tubes, no header at all is needed. The inlet tubes could also extend into the storage tank and join a header that resides inside this tank.

The Bio-Energy Systems, Inc., system is excellently suited to the construction of air-type collectors. Here the absorber mat and headers are dispensed with, and a simple black sheet of aluminum, say, is used as absorber. Or one can employ a foil-faced insulation panel, such as Thermax, and paint the foil black.

The system is well suited to various domestic-hot-water systems.

The system performs well in water-type thermosiphon systems. Here the tubes run vertically, instead of horizontally; the direction of flow is the same in all tubes, and the headers are situated at top and bottom.

The system is excellently suited to retrofit application.

The rubber framing strip can be used for installing additional glazing on living room windows, bedroom windows, basement windows, and skylights. It is applicable also to greenhouses. It can be used with curving window frames and curving roofs or walls.

The rubber absorber mat can be employed in radiant heating: it can be installed beneath a layer of concrete, or it can be incorporated in a ceiling or wall. Because the mat can be bent 180° easily, and comes in lengths up to 600 ft., installation is rapid. Because the mat is flexible, no allowance for thermal expansion is needed. There is no danger of corrosion. The mat can be used also for various heat-exchange purposes: it can be wrapped around a steel tank or looped around inside the tank; thus, it is well suited for use in domestic–hot–water preheating. It can be used also for various cooling purposes. It can be used to dissipate heat in a nighttime radiative cooling system, or to dissipate heat from an air–conditioner or heat–pump that is cooling a building. It can be used with many kinds of liquid, including brine.

DISTRIBUTORS

The components are distributed, not by the manufacturer, but by regional agents.

BELOW-ATMOSPHERIC-PRESSURE COLLECTOR
OF D. L. SPENCER ET AL.

SUMMARY

This water–type collector operates at below–atmospheric pressure, with six helpful consequences: higher efficiency, lower cost of materials, faster warm–up, elimination of harm from freezing or boiling, mechanical flexibility.

INTRODUCTION

The idea of developing a water–type collector that would operate at slightly below atmospheric pressure was conceived several years ago by D. L. Spencer of the University of Iowa. Experimental models of such a collector have been built by him and by Pleiad Industries, Inc., of West Branch, Iowa. Recent tests have been successful: the collectors tried out do indeed exhibit the favorable characteristics predicted.

DETAILS

In this collector, water flows downward throughout the entire area of the collector. There are no tubes. The entire back side of the black, radiation-absorbing sheet is in contact with flowing water. The water is confined between two sheets of stainless steel, each being 0.006–in. thick. The upper sheet is blackened. The lower sheet is textured, that is, it has a finely grooved or corrugated shape, the grooves running up and down the sheet. The grooves are about 0.015–in. deep. The upper and lower sheets are pressed against one another with a force of the order of 1 psi, because the absolute pressure in the intervening water is 1 psi below atmo-

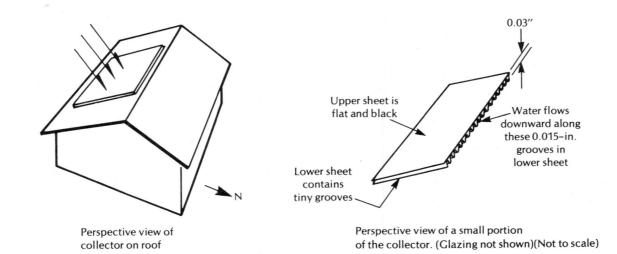

Perspective view of
collector on roof

Upper sheet is
flat and black

Lower sheet
contains
tiny grooves

0.03″

Water flows
downward along
these 0.015–in.
grooves in
lower sheet

Perspective view of a small portion
of the collector. (Glazing not shown)(Not to scale)

spheric pressure. Despite the pressure of each sheet on the other, the water continues to flow—in the tiny grooves. The pathlength for heat flow (from black surface to water) is 0.006 in.; accordingly the delta-T penalty is vanishingly small. The amount of water and metal in the collector is so small that the warm-up time is very short. Conventional glazing, of fiberglass and polyester, for example, is used. Below the collector proper there is a layer of insulating material. The desired below-atmospheric pressure can be maintained by any of several simple methods. One elegant method maintains the same pressure all the way along the collector, from top to bottom: the downward force of gravity is everywhere balanced against the upward force of friction, provided that the proper flowrate is maintained.

HELPFUL CONSEQUENCES

The entire underside of the black absorbing sheet is swept by coolant, practically eliminating delta-T loss.

Because the two sheets are pressed together, they have no tendency to bulge. Accordingly, they can be very thin. Accordingly, they are cheap—even stainless steel, in this thickness, is cheap; they are easy to lift and transport; they are flexible (see below); their thermal capacity is small.

There are no tubes within the collector—no slender tubes to become blocked. If some local blockage does occur, the water can simply flow around it.

If a small amount of freeze-up occurs, the upper sheet is free to bulge. It does not burst. When the ice melts the sheet returns to its normal shape.

If a small amount of boiling occurs (in summer, say), bulging relieves the stress.

Most parts are flexible and can be bent to conform to a wavy, or slope-changing, roof. The system is virtually "drapable."

DRAWBACKS

Equipment for maintaining the proper below-atmospheric pressure must be provided. The edges of the collector must be sealed carefully. The system should not be allowed to become so hot that prolonged boiling occurs.

References

"Design and Performance of a Distributed-Flow Water-Cooled Solar Collector" by D. L. Spencer, T. F. Smith, and H. R. Flindt of the Division of Energy Engineering, University of Iowa, Iowa City, IA 52242, 1975. Also an article by D. L. Spencer in *Proceedings of the American ISES Aug. 1978 Conference*, Denver, CO, Vol. 2.1, p. 629.

KHANH'S RADICALLY NEW APPROACH TO INCREASING THE USEFUL OUTPUT OF A FLAT-PLATE COLLECTOR PANEL: SCHEME EMPLOYING A CHEAP, LATERALLY CONTIGUOUS, SLAVE PANEL

11/8/76

SUMMARY

D. Khanh has invented a radically new way of approximately eliminating, in effect, the conductive and convective heat losses of any given flat–plate collector panel.* The method, which applies only if the available region of irradiation (roof area, say) considerably exceeds the area of the given panel, is to provide the given panel (*master* panel) with a laterally contiguous, cheap, air-type panel (slave panel), which serves by an overreaching gravity–convective circulation of hot air to keep the upper (outer) part of the master panel much hotter than it normally would be. See Fig. 1. With respect to absorption of solar radiation, slave panel and master panel are in parallel; but with respect to losses by conduction and convection, they are in series, with the slave uppermost. Thus the slave "pays the bill" on behalf of the master. The sole function of the slave is to supply the energy that (unavoidably) is lost from the master's outer glazing layer. The master itself is then, in effect, practically lossless. Practically all of the energy it receives from the sun is delivered to the storage system; in fact, under especially favorable circumstances the amount delivered can exceed the amount received by the master from the sun.

*Reports and letters received in Sept. and Oct. 1976 from Dinh Khanh of 2221 NE 12th Terrace, Gainesville, FL 32601. Mr. Khanh is associated with Mathews Systems, Inc., PO Box 1666, Gainesville, FL 32601.

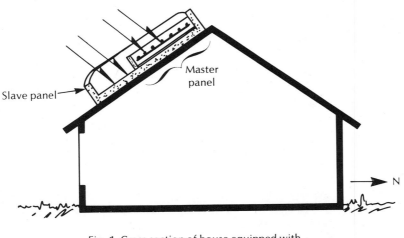

Fig. 1. Cross section of house equipped with master panel and slave panel, the latter's airspace extending over the former

Because the slave can be of very crude design and construction, it can be very cheap. Because it is so cheap, the system as a whole may deliver more Btu's per buck than the master alone could deliver.

The scheme is most advantageous if the master panel is especially expensive and—lacking a slave—would normally be operated at such high temperature that the energy losses would be large.

Khanh reports that actual tests have shown the effectiveness of his scheme.

Note: In some special situations (discussed at the end of this report), it may be better to employ a flanking crude reflector than to employ Khanh's slave panel.

INTRODUCTION

In 1976 I learned from Dinh Khanh of 2221 NE 12th Terrace, Gainesville, FL 32601, of a radically new principle for increasing the useful output of a flat–plate solar-radiation collector-panel of almost any sort. In many circumstances the useful output can be increased by about 25% to 35% (I guess) and in some especially favorable circumstances the increase may be much larger. The scheme is especially applicable to (a) collectors that have high cost per square foot and (b) are normally operated in such a way that the losses are large.

Here I try to explain the principle in a somewhat different way from that which Khanh has used. I adopt an elementary and stepwise approach.

PRELIMINARY REMARKS CONCERNING SOME OLD CHOICES

Let me first dispose of some old choices and avenues—avenues *not* used by Khanh.

A well-known choice is between (1) a costly, high-performance collector, and (2) a cheap, low-performance collector. Which is better? Whichever provides the "most Btu's per buck," if durability, visual appearance, etc., are equally good.

Another well-known choice is between (1) single glazing and (2) double glazing. With double glazing the conductive and convective losses are smaller, but the reflection loss is greater and the cost is greater.

Other choices involve selective coatings, honeycombs, infra-red-reflective coatings, etc.

Khanh's approach is radically different.

KHANH'S PRINCIPLE

The heart of Khanh's principle is the use of a cheap collector-panel and an expensive collector-panel side-by-side, with the former "downhill" from the latter and serving the latter by canceling its losses.

More exactly, the new principle calls for use of a cheap panel that plays a small role and an expensive panel that plays a large role. The cheap panel (slave) serves solely to assist the expensive panel (master). With respect to absorption of solar radiation, slave and master are in parallel. With respect to loss of heat to the cold ambient air, slave and master are in series, the slave extending over the master and bearing the brunt of contact with the cold ambient air.

Figures 2–5 explain the principle.

Figure 2 shows a conventional, water-type collector-panel (master). It is double-glazed with glass. It includes a black copper sheet and black copper tubes. Water is circulated in the tubes and carries energy to the storage system.

Figure 3 shows the same master panel and, beside its lower edge, a crude (slave) panel. The slave is double glazed with plastic films. It contains nothing except air—no copper sheet, no tubes, no water. Its lower face consists of a black sheet of aluminum foil, fiberglass backed. No coolant enters or leaves the slave. The slave serves, here, no useful purpose. At noon on a sunny day the stagnant air in the slave is extremely hot.

Figure 4 again shows master and slave; but here the hot air from the slave is free to circulate, by gravity convection, between the two layers of glazing of the master. The circulating hot air keeps the master's outer glazing layer moderately hot and keeps the inner layer very hot—so hot that the black copper sheet loses little or no heat upward by conduction or convection. (If the airspace above the master's inner glazing is too thin, the hot air from the slave cannot circulate here very rapidly by gravity convection; the airspace must be fairly thick, I suppose.)

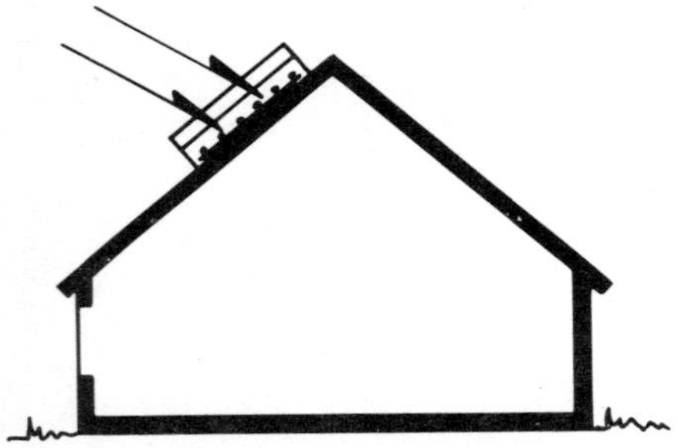

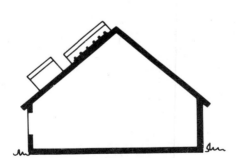

Fig. 2. Master panel
on south roof. (Not
to scale)

Fig. 3. Master panel and
slave panel on roof.
There is no communication
between the two.

Fig. 4. Master panel and
communicating slave panel

Note: In the drawings the water-filled tubes run horizontally. They were drawn this way to make the drawings more understandable. In practice most designers would arrange, presumably, to have the tubes run up and down the roof, as usual.

Figure 5 shows the pair of panels in greater detail. Small arrows indicate the circulation of air from the slave.

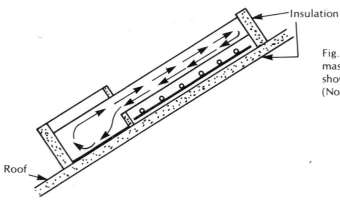

Insulation

Roof

Fig. 5. Enlarged view of master and communicating slave, showing directions of airflow (Not to scale)

Figure 6 shows a much cruder slave—it has no black metallic absorber—the black roof surface itself serves as absorber. Also, the slave has practically no frame—the glazing is attached to the master and to the roof itself. Such slave panel might cost, I guess, only about $2. per square foot, installed.

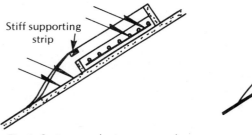

Stiff supporting strip

Fig. 6. System employing very crude, cheap slave

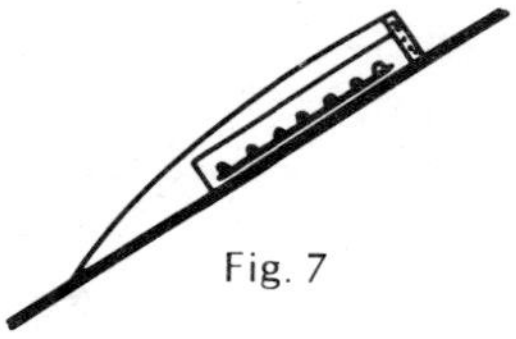

Fig. 7

Fig. 8

Even cheaper designs may be devised. The slave may have single glazing only. Perhaps its (single) glazing can be a mere extension of the outer glazing sheet of the master. In fact the entire slave may be regarded as an edge-frill on the master and may cost an order of magnitude less, per square foot, than the master costs. See Figs. 7 and 8.

Circulation of hot air from the slave might be facilitated by allowing this air to return *beneath* the master. See Fig. 9.

One might use a master panel that is single glazed with glass (which provides good greenhouse effect) and a slave that is double glazed with plastic (plastic being cheap and easy to ship and install).

One might extend the hierarchy to comprise master, slave, and subslave. See Fig. 10. Here the master could collect much energy even at, say, 200°F.

With the aid of a computer, one might find out fairly precisely which scheme is most cost-effective. I find it difficult to guess. For different circumstances, different designs may be optimum.

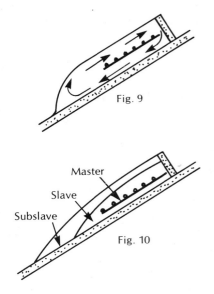

Fig. 9

Fig. 10

Discussion of Increase in Output of Useful Energy

Approximately 100% of the energy lost to the outdoors is supplied by the slave, and approximately 100% of the energy received by the black copper is delivered to the storage system.

If, without the slave, the master panel would deliver 70% of the absorbed energy to the storage system (losing 30% by various loss mechanisms), *with* the slave about 100% of the absorbed energy would be delivered to storage. This represents an improvement of $(100 - 70)/70 = 30/70 = 43\%$.

In periods of lower level irradiation or periods with intermittent irradiation, the benefit produced by the slave would be less in absolute terms but greater in relative terms. Under such unfavorable conditions, the master might normally deliver only 40% of the absorbed energy to the storage system, but, with the slave helping, might deliver (a guess) 70%. This is an improvement of about $(70 - 40)/40 = 75\%$.

A very favorable fact is that the slave, with its near-massless absorber (black aluminum foil, fiberglass backed) and its near-massless glazing (plastic film) has an extremely short warm-up time. Thus, it springs into action fast, at start of day or when the sun comes out from behind the clouds. Besides performing well under steady-state, sunny-noon conditions, it in effect lengthens the day. Effective delivery of energy to the storage system starts earlier and continues later.

Under some very special circumstances the amount of energy delivered to storage can exceed the amount of solar radiation reaching the black copper sheet. This would happen if the slave were large in area, the master were small in area, and the average temperature of the circulating water were very low.

Discussion of Applicability

I expect that the main use of the slave will be in assisting collection at high temperature, such as 160°F or 210°F or higher temperatures. If the storage system is heated to 180°F instead of 140°F, its carrythrough is much greater; or one might use a smaller storage system and yet achieve the originally contemplated carrythrough. If the storage system is heated to 210°F, the operation of a LiBr absorption chiller becomes highly efficient, relative to use of water at 190°F.

Khanh's principle is applicable to water-type collectors or air-type collectors. If water-type collectors customarily operate at higher temperature than air-type collectors and are more expensive than air-type collectors, then the principle may be applied especially advantageously to water-type collectors.

The principle is applicable whether the master panel's slope is of intermediate value (such as 30° to 60° from the horizontal) or greater or less. It is applicable even if the master is vertical. It may be workable even if the master is horizontal.

Probably the scheme is applicable even to panel arrays that are long and wide. One could arrange several rows of master panels, with wide spaces between them for slaves, as suggested by the following diagram. Conceivably one could use an array of columns: spaced columns of masters with slaves between. (One could even use a small blower to circulate hot air from a slave to a master. It would then be feasible to use thinner airspaces and greater panel widths.)

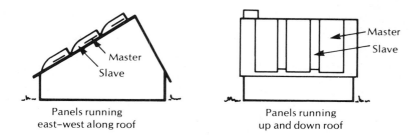

Panels running
east-west along roof

Panels running
up and down roof

WHY NOT MERELY USE A LARGER AREA OF CONVENTIONAL PANELS?

If someone proposes that, instead of using slaves, he use a larger area of conventional water-type master panels, he should bear in mind the full incremental cost of the added master panels. These costs include:

    purchasing the additional panels

    shipping them

    hoisting them onto the roof and securing them

    connecting up the coolant tubes

    providing an additional amount of water-and-glycol

    increasing the length and diameter of the supply pipes

    increasing the power of the centrifugal pump

Someone might propose using, in place of the master and slave, an all-air collection system. He might expect to enjoy the economy of slave-type construction without suffering the expense of the water-type panels. But he should remember that, using a conventional air-type system, he must use main ducts and feeder ducts; the main ducts must extend down into (and along) the basement; the ducts take up much room; they must have no air-leaks; they must be well insulated on all sides; a large blower is required; a large bin of stones must be used or a large air-to-water heat-exchanger may be used to transfer the heat to a tank of water. Khanh's scheme uses no ducts, no blowers. It takes up no room within the house. His slave is small, cheap, passive, connected to nothing.

SLAVE PANEL VS. FLANKING REFLECTOR

This question arises: Why not use, instead of a contiguous slave, a flanking reflector, such as is sketched here? The article by McDaniels et al. in the Nov. 1975 *Solar Energy* explains why a flanking reflector, such as is used in the Matthew House in Coos Bay, Oregon, can almost double the amount of energy delivered to the storage system.

Each of these assisting devices—slave panel and reflector—takes advantages of extra space off to the side. Each is cheap. Each springs into action fast. Each effectively lengthens the day.

If the master panel is vertical or close to vertical, I tend to favor use of a reflector adjacent to the lower edge of the master. But if the master is 50° to 70° from the horizontal, the geometry is unfavorable to the use of reflectors. Here I favor use of the slave panel.

If the master is horizontal, or tilted up to, say, 30° from the horizontal, there may be merit in using both kinds of flankers: the slave near the lower edge of the master and the reflector near the upper edge. This arrangement might become of importance for use in powering LiBr absorption chillers in, say, the southern half of the US; here, the master would normally be tilted only about 30° from the horizontal, in order that much energy could be collected in summer.

For summer

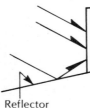

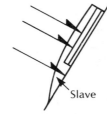

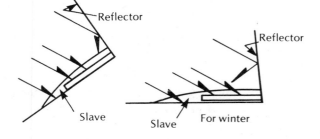

For winter

The tilt of a reflector can be adjusted—in spring and fall, say. Also, the reflector may be swung to cover the master to protect it from snow or vandals or to reduce heat loss at night. But, in some environments, a simple aluminum reflector may soon become dull.

FINAL WARNING

I hope I have correctly explained Khanh's main idea. But I may have misunderstood some aspects of it or given incorrect emphasis to various aspects. Also, I have included, especially toward the end, some related ideas of my own, and if I have made mistakes in presenting these ideas, no blame should accrue to Khanh.

ALL-AROUND-THE-PERIPHERY, ACTIVE, AIR-TYPE COLLECTOR FOR A SMALL, SOUTH-FACING WINDOW

SUMMARY

A small (3 ft. × 2-ft.) south-facing window of a south room allows some solar energy to enter the room. Is there a simple method of greatly increasing the amount of solar energy entering via such window? Solar Room Co. of Taos, New Mexico, has developed a method; their equipment is called a *window expander*. Not knowing enough about that company's equipment to be able to describe it with confidence, I describe here my own ideas as to how the goal might be accomplished. But I expect that the Solar Room Co. scheme is better.

I would install a 7 ft. × 7-ft. transparent plastic sheet just outside the window and symmetrical with respect to it, i.e., extending over all areas on all four sides of the window. The sheet would be sealed along the edges. The lower sash of the window would be raised a few inches and two ports would be provided in a board that fills the few-inch opening. On sunny days the warm air in the space defined by the plastic sheet would be circulated to the room by a small blower.

A modified scheme that is entirely passive might be used.

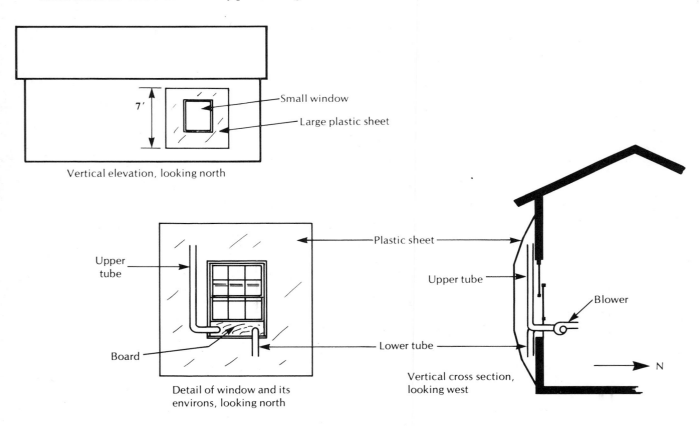

Vertical elevation, looking north

7'

Small window

Large plastic sheet

Upper tube

Board

Detail of window and its environs, looking north

Plastic sheet

Upper tube

Lower tube

Blower

N

Vertical cross section, looking west

INTRODUCTION

There are, of course, various devices that can increase the amount of solar energy that enters a house via a small south-facing window. For example, a near-horizontal reflector installed outdoors near the base of the window will reflect additional radiation toward it. A sloping, air-type, thermosiphon collector box installed outside the window, and below it, will deliver hot air to the room.

Engineers at Solar Room Co., Box 1377, Taos, NM 87571, have invented a scheme, called a window expander, that represents a different approach. They employ a large rectangular enclosure that is installed approximately symmetrically with respect to the window. The scheme described below is one invented by me. It has some resemblance to their scheme, I understand.

PROPOSED SCHEME

Install, just outside the small vertical south window, a thin transparent plastic sheet that is approximately symmetrically situated with respect to the window and is about 4 to 8 inches from the face of the window and south wall. The plastic sheet, which is 7 ft. × 7 ft., is slightly bowed, or curved; slender ribs of wood or metal maintain the curve. The purpose of bowing the sheet is to prevent it from flapping in the wind. The enclosed portion of the south face of the house is painted black, or other dark color, so as to absorb most of the incident solar radiation. The radiation heats the dark-painted surface and this in turn heats the air that is within the 7 ft. × 7-ft. space or chamber. This air tends to rise; the hottest part of the chamber is the uppermost part.

The lower sash of the window is raised 5 inches and a 2-in.-thick wooden board is inserted snugly in the 5-in. space. There are two 3-in.-dia. holes in this board. One serves a plastic outdoor duct, or tube, that extends 1 ft. downward into the chamber and permits flow of room air into the chamber. The other serves a similar tube that extends upward almost to the top of the chamber; a small blower draws hot air from this tube and delivers the air to the room.

The blower runs whenever a sensor finds that the air in the top of the chamber is hotter than 80°F. Whenever the blower stops, very-lightweight dampers (of 0.001-in. polyethylene) close due to the pull of gravity and prevent flow of air to or from the chamber.

On a sunny day, the chamber, with an area of almost 50 ft.^{2}, may deliver 6000 Btu per hour to the room, i.e., almost 2 kW of heat, equivalent to 10¢ worth of electrical energy. This corresponds to 50¢ worth per sunny day, 25¢ worth per typical day, or about $40 worth per winter. If such a system, purchased as a kit shipped by US mail or UPS and installed by the homeowner, were to cost $80, it would pay for itself in 2 years relative to the use of electrical energy and in 6 years relative to use of fuel oil.

DISCUSSION

The scheme does not entail cutting any hole in the wall of the house. Most of the window area is scarcely affected; it still admits sunlight. The entire equipment can be removed at the end of winter and stored in a garage. If a horizontal reflector is installed outdoors near the base of the chamber, the thermal output by the chamber will be increased considerably.

The system developed by Solar Room Co. is, I understand, somewhat similar. In some embodiments of its system, two or three sheets of plastic are used, with the result that loss of heat to the outdoors is reduced. The sheets are of 0.006-in. polyethylene, and the inner ones may be joined so as to form an array of cavities or ducts. At night, with the blowers off, the cavities deflate and lie close to the window and wall areas and serve as a thermal curtain. In applications to commercial buildings, office buildings, and other buildings that have massive south walls, emphasis is placed on storing much of the solar energy directly in these walls; thus, overheating of the rooms during sunny hours is avoided and the walls help heat the rooms at night.

MODIFICATIONS

Scheme S-91a

Same as above except use double glazing such as has been used for some years by Solar Room Co., in its solar walk-in enclosures. A very small blower maintains air pressure that keeps the two glazing sheets 1 or 2 inches apart. One blower could serve several different chamber-equipped windows on the given south wall. Or it could serve a single very large chamber, say 7 ft. high and 36 ft. long.

Scheme S-91b

Here the system is entirely passive. To facilitate thermosiphon flow, larger diameter tubes (ducts) are used. One tube is associated with a board installed below the lower sash and the other tube is associated with a board installed above the upper sash. This latter tube extends upward within the room to provide chimney-effect assistance to the thermosiphon flow. Again, passively actuated dampers are used; on cold nights they may be supplemented by sturdier, manually operated dampers.

The system requires no sensor, no blower, and no electric power. It performs well even if the electric power supply fails. (But will the flow be sluggish? Is it acceptable to have the hot air delivered to a region very near the ceiling?)

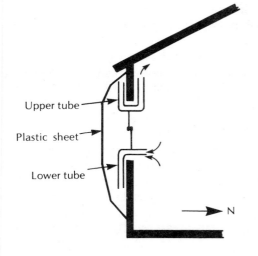

Passive thermosiphon system

Scheme S-91c

This passive system is like the one described above except that only one board—the lower one—is used. Both tubes pass through holes in this board. How can one arrange for thermosiphon circulation to occur through these two holes, which are at the same level? By arranging to have the tube for air flowing from chamber to room to extend far upward in the room to provide chimney effect. This tube, which is of translucent plastic, passes upward close to the window, i.e., so as to intercept some solar radiation that has passed through it. This portion of the tube contains a central black strip of aluminum foil (or cloth or plastic), which absorbs solar radiation, becomes hot, and warms the air here causing it to rise. In other words, there is a passive, solar-powered forced draft. (An alternative procedure, or supplementary procedure, would be to cut a small hole in the lower portion of this tube that is in the special chamber; thus a little warm air from window-sill level in the chamber can escape into the indoor "chimney" and initiate upward flow there. Once the flow is started, it will accelerate and draw air via the upper part of the tube in the chamber. The small hole serves as "bleeder hole" to get the main thermosiphon flow started.)

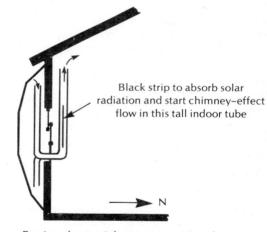

Black strip to absorb solar radiation and start chimney-effect flow in this tall indoor tube

N

Passive, thermosiphon system using only one board. Both tubes pass below the lower sash.

PENTHOUSE AIR-TYPE SYSTEM EMPLOYING SHEETLIKE AIRFLOW PRODUCED BY A MULTI-FUNCTION BLACK ABSORBING MATTRESS THAT SERVES AS ABSORBER, BELLOWS, AND [AT NIGHT] THERMAL SHUTTER

Scheme S-138
9/30/76

SUMMARY

Here I discuss only the active portion of a solar system (of a two-story house) that includes a passive system and an independent active system. Either system alone might provide, say, 55% of the winter's heat-need. Together they provide 95% (guess). The passive system is conventional, making use of the vertical south windows of both stories of the house.

The active system, of air-type, is of unique design. The collector and storage systems are situated cheek-by-jowl in a slender, 40-ft-long penthouse on the north portion of the roof. The collector is vertical and single-glazed. The storage system includes 16,000 lb. of water in one-gal. bottles. South of the penthouse there is a reflecting (aluminized) roof area that increases collector energy-receipt by 40% and increases energy-delivered-to-storage by 80%. The 36-ft-long absorbing sheet, spaced 2 in. from the glazing, is part of a 4-in.-thick, dynamic, multi-purpose septum that includes 3½ in. of insulation. The septum constitutes the moving element of a 36-ft.-long "bellows" that produces simple, sheetlike flow of air upward in the collector and downward in the contiguous bin—without need for any ducts. Total pathlength of flow is small—13 ft. Total cross-section of flow is large: 6 ft². Solenoids drive the septum every two seconds, cyclically—silently and efficiently. At night the (thick insulating) septum is pushed against the glazing; thus, at night the penthouse is heavily insulated on *all* sides. To distribute heat from the storage bin to the rooms, small cheap ducts and blowers are used.

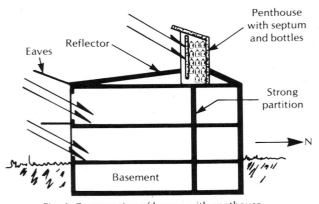

Fig. 1. Cross section of house, with penthouse, reflective roof area, huge eaves

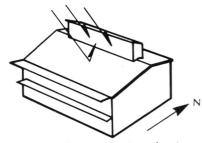

Fig. 2. Perspective view, showing that the penthouse is scarcely visible to persons on the ground

The proposed system avoids the need for huge, well-insulated ducts and large, noisy blowers. It makes no use of the basement. The penthouse is almost hidden from view from the ground. It is accessible, as for servicing, from all sides. Various alternatives are suggested.

DETAILS

I assume a two-story New England house that is 54 ft. in E–W dimension, has huge south double-glazed windows, and has massive floors and walls. Exterior walls are heavily insulated on the outside. The main part of the roof slopes 6° upward toward the north.

Passive solar heating: This is accomplished with the aid of the large south windows and the massive walls and floors. The design is similar to that of Saunders (or that of Anderson, or Wright, or Lasar, or others), and will not be discussed here. The system could, by itself, provide 55%, say, of the winter's heat need. Eaves and balcony shade the south windows in summer.

Active solar heating: This is accomplished with the aid of a special penthouse situated 10 ft. from the north edge of the roof and directly above a strong, E–W partition wall. The penthouse is 40 ft. long, 7 ft. high, and 4 ft. in N–S dimension. A 54 ft. × 12-ft. roof area just south of the penthouse is aluminum faced to direct more radiation toward the penthouse. The S face of the penthouse is single-glazed with non-absorbing glass. The N face consists of panels (access panels, removable with the aid of a screwdriver) that are insulated with 8 in. of fiberglass and well sealed at the edges. The ends and top of the penthouse are similarly insulated.

About 2 inches from the glazing there is a multi-purpose septum, 36 ft. × 6 ft. × 4 in., which includes a thin black sheet of aluminum, 3½ in. of insulation, and a flexible backing sheet with horizontal stiffening bars near top and bottom.

The space north of the septum is nearly filled with a loosely packed array of 2000 1-gal., water-filled bottles. Total mass of water: 16,000 lb.

Air is forced to circulate upward in sheetlike flow, in the above-mentioned 2-in. space, and over the top of the septum, then down through the array of bottles and thence back to the base of the 2-in. space. Total length of airflow circuit: about 13 ft. Cross-sectional area of flow in 2-in. space: 2 in. × 36 ft., i.e., 6 ft.^{2}. Total number of corners in the airflow circuit: two—and these can be rounded.

The sheetlike flow is maintained without benefit of blower or ducts. Rather, the air is driven by the reciprocating and undulating motion of the septum, which, operating in opposition to the vertical glass sheet, constitutes a kind of giant bellows. The septum is suspended freely and is somewhat flexible. At 2-sec. intervals, its lower portion is pushed 2 in. southward (until it presses against the glass sheet) by means of electromagnetic solenoidal actuators, and 0.7 sec. later its upper portion is similarly driven southward. The solenoids push against the above-mentioned horizontal stiffening bars; these are located slightly below the centers of mass of the respective septum portions, which enhances the constructive undulating character of the motion, which in turn drives the air upward—not equally upward and downward. Operation is smooth and efficient. During the recovery strokes of the solenoids, the septum swings back to its normal position as a result of gravity-and-pendulum effect; the bottom of the septum recovers first, and the top recovers 0.7 sec. later; accordingly, relatively cool air from the base of the array of bottles flows into the 2-in. space that opens up near the bottom of the septum, and soon this air in turn is driven upward. Thus the cycle repeats, and the up-and-over flow of air within the penthouse as a whole continues. (Note: I know almost nothing about solenoids. Perhaps they should not be used. An alternative propulsion means is discussed on a later page.)

Figures 3 and 4 show details of the septum, array of bottles, etc.

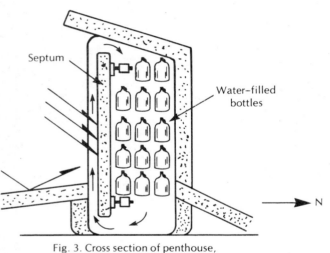

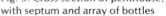

Fig. 3. Cross section of penthouse, with septum and array of bottles

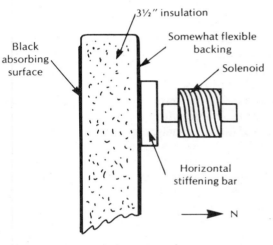

Fig. 4. Detail of septum and driving device

Figure 5 shows various phases in the cycle of septum motion. For simplicity, we have omitted the horizontal stiffening bars and the solenoids.

When, at night or on sunless day, the rooms threaten to become cold and the heat stored in the walls and floors no longer suffices, heat is drawn from the array of bottles in the penthouse. It is drawn by means of 6-in.-dia., flexible, lightly insulated ducts, housed informally within the framing of the house, and fractional-HP blowers. I do not bother to discuss this system in detail. The required rate of delivery of heat is an order of magnitude less than the rate at which heat is taken in by the penthouse on a sunny day; thus, a relatively trivial set of ducts and blowers suffices.

When the sun is not shining, the entire septum is swung close against the vertical glass sheet, virtually stopping heat loss through the south face of penthouse. It may be swung by any of various available means, which I shall not discuss here.

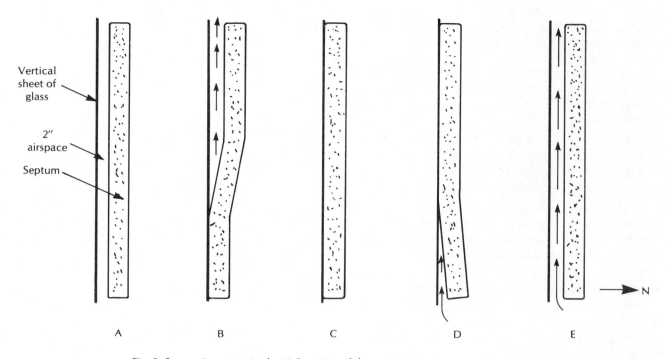

Fig. 5. Successive stages in the N–S motion of the septum
A.   Start of cycle—septum is 2 in. from vertical sheet of glass.
B.   Lower portion of septum is pressed southward (to left) against glass.
C.   Upper portion likewise has been pressed southward.
D.   Lower portion of septum has relaxed to normal position, drawing in air from below.
E.   Upper portion likewise has relaxed. Cycle now repeats.

DISCUSSION

The proposed scheme is bold in several ways:

1.   It allows full use of the south face of a house for (simple, reliable, almost free) passive solar heating. The active system is *in addition to* a full-size passive system. The two systems complement each other. Neither encroaches on the other.

2.   It places the active collector in an out-of-the-way location almost invisible from the ground, inaccessible to vandals.

3.   It avoids several major headaches of ordinary air-type systems:

     It requires no huge-cross-section ducts that, in aggregate, may be over 100 ft. long (compare with CSU Solar House II); no large ducts that must be heavily insulated; no ducts that must be leak-tight) no large ducts that must turn sharp corners. No ducts at all are involved in the collection process.

It avoids placing the collector and storage bin "poles apart"—one on top of the house, the other in the basement. It places them together.

It avoids the need to repeatedly "neck down" the airflow and then spread it out again. The flow does not have to be necked-down to pass through a highly localized blower; it does not have to be spread out again in a bin-of-stones; it does not have to be spread out again to pass through the collector. In the proposed collection scheme, no necking-down is involved at all: the airflow is at all times sheetlike; at all times the width of the stream is 36 ft.

It avoids the efficiency losses inherent in the operation of a rotary blower, which produces (locally) extremely high-speed flow and an extreme degree of turbulence.

It avoids the noise associated with a ½-HP or 1-HP blower.

It avoids the embarrassment of having approximately the *same* rate of airflow during the collection process and during the distribution of heat to the rooms (when, in fact, a 5-or-10-fold lower rate of airflow is adequate for the latter).

It leaves the basement unencumbered: the basement contains no ducts, no blower, no storage bin. The entire basement is available for use as a recreation area or for storage, etc.

It provides service walk-way areas on both sides of the roof-top equipment. (In most air-type solar-heated houses, access to the collector is difficult.)

Some Additional Favorable Features of the proposed scheme are:

The passive system heats, mainly, the south part of the house. The active system, situated over the *north* part, is well suited to heating this part.

Because the thick, insulating septum can be pressed against the vertical glass window in sunless periods, little heat-loss occurs through that window, and accordingly it is permissible that this window be single-glazed. Being single glazed, it transmits about 10% more radiation than a double-glazed window would transmit.

The collector geometry is well matched to that of the crude reflector just south of it; the criteria stated by McDaniels *et al.* (*Solar Energy*, Nov. 1975) are satisfied: the angle between reflector and window is close to 90°; the reflecting region is twice as wide as the collector window is high; the reflecting region extends beyond the ends of the penthouse. One may expect the reflector to add about 40% to the amount of energy reaching the black absorbing surface and to add about 80% to the amount of energy delivered to storage.

The aggregate area of the bottles is ample: several times the area of the black absorbing surface of the collector.

Thermal capacity of collector proper is very small. Warm-up time—only 2 (?) minutes.

MODIFICATIONS

One could install a large coil of copper pipe (or a small tank) in the penthouse and use this to preheat domestic hot water.

On cool summer nights one could circulate cool outdoor air through the array of bottles, cooling them. During hot days, room air could be circulated through the (cool) array. Note: Because, normally, air flows through, say, the east and west halves of the array of bottles in parallel, one could, in summer, keep the east part hot and keep the west part cool. Thus, the penthouse could serve two opposite functions simultaneously—heating domestic hot water and cooling the rooms.

One could install three 100-watt electric heating elements in lower half of the penthouse and have these turned on automatically when and if the temperature in the penthouse falls below 45°F. Thus risk of freeze-up could be avoided.

One could make the penthouse wider and provide a service walkway along the center.

Instead of using solenoids to drive the septum south, toward the window, one could use the reverse scheme—employ springs (or gravity) to move the septum to the south and employ the solenoids to move the septum to the north. Then, if the electric power supply fails, the septum reverts automatically to its fail-saft (insulating) position. Also, no electrical power is needed to *hold* the septum in the south position.

Instead of using solenoids to make the septum swing back and forth, use a small electric motor (bolted to the septum) that has a gear reduction system and a 30-rpm output shaft on which a heavy weight is mounted eccentrically. Because of the eccentric rotating weight, the entire septum will "wobble" north and south.

IMPORTANT QUESTION

Should one do away with the septum entirely and let the solar radiation proceed directly to the bottles? The answer is: If one did, one would then face the problem of preventing the bottles from losing heat all night through the collector window. To prevent such heat-loss might be expensive. See, however, Scheme S-139.

PART 4

Concentrating Systems

INTRODUCTION

The variety of concentrating collection systems developed in recent years by groups here and abroad is truly enormous. I have made a list of about one hundred kinds of systems—but the list is too long, too complex, and too incomplete to include here.

Presented below is a list of the main categories—and accounts of some proposed new schemes.

Classes of Concentrating Systems
Concentrating collectors may be classified in many ways, e.g.:

By *physical method* of changing the directions of the sun's rays
1. Refraction—use of lenses
2. Reflections—use of mirrors

By *geometrical form or shape* of the direction-changing device
1. Portion of a sphere—spherical lens or mirror
2. Portion of a cylinder—cylindrical lens or mirror
3. Curved array of flat strips

By *extent of subdividing* the direction-changing device
1. No subdivision—simple lens or mirror
2. Much subdivision—fresnel lens or mirror
    (named after the French scientist A. J. Fresnel who made important discoveries in optics in about 1815).

By *extent of tracking* used
1. No tracking
2. Periodic re-aiming—weekly or monthly adjustment
3. Full, continual tracking

Most of the schemes we describe here involve cylindrical reflector or cylindrical fresnel lenses. Obviously, it is not necessary

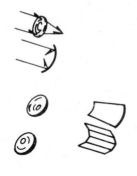

that the reflectors or lenses be precision-made: they are not used to produce sharp images but merely to collect radiation. Thus, they can be of cheap design and construction. In several of the schemes the reflector serves also as a thermal shutter.

MULTI-PURPOSE, WINDOW-SEAT-TYPE, PASSIVE COLLECTION-AND-STORAGE SYSTEM THAT INCLUDES THREE-MODE, HINGED, REFLECTING-AND-INSULATING PLATES

Scheme S–177t
8/5/77

PROPOSED SCHEME

The heart of the system is a side-by-side pair of long, slender, cylindrical, horizontal, water-filled, galvanized steel tanks. Each is 14 in. in diameter and has a selective black coating. The pair is enclosed in a 3-ft.-wide, 1½-ft.-high housing situated on the floor and close to the vertical, 8-ft.-high, double-glazed south windows. The sides and bottom of the housing include 3 inches of insulation and there are channels between insulation and tanks to permit circulation of room air (circulation that may be assisted by a small fan). Close above the pair of tanks there is a sheet of tough plastic glazing.

Above the pair of tanks there is a 6-ft.-high, cylindrical, approximately parabolic, crude reflector (employing, say, King-Lux aluminum or aluminized mylar) which, throughout about a 5-hour midday period directs much incoming solar radiation downward toward the pair of tanks. The reflector backing includes 3 inches of insulation. The reflector consists of two halves. The *upper* half H_1, when not needed to direct radiation toward the tanks, can be (1) manually swung toward the window, to act as thermal shutter for the upper half of the window—at night or on very dark days, or (2) manually swung toward the ceiling (and held there by a hook) to be out of the way and allow much light to penetrate deep into the room and to allow a view. Attached to the back of H_1 there is a hinged "rider panel" which (when H_1 is in position close to the window) can be swung down to insulate the lower region of the window. The *lower* half, H_1, which is very strong, can be manually swung downward toward the window, thus, (1) providing an insulating cover for the housing and (b) providing a place to sit or lie (like a window seat or couch).

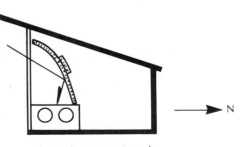

Vertical cross section of house, looking west

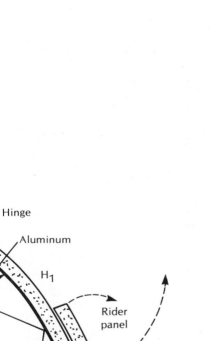

Vertical cross section of solar heating system, looking west

Domestic hot water is preheated while flowing along two pipes just above the tanks, and off-peak electric heating may be applied to the storage system by means of some electrical heating strips just below the tanks.

OPERATION

On cold sunny mornings H_1 is swung upward and H_2 downward, to allow solar radiation to penetrate deep into the room. However, during most of the sunny day H_1 and H_2 are oriented to direct much radiation toward the tanks. At night, H_1 and its rider are used to prevent heat-loss through the window; H_2 is in down position. Vents in the north side of the housing are opened to allow room air to circulate past the tanks, to carry heat to the room. A small fan can be used to speed this process.

If the tank system is 30 ft. long, it holds 4000 lb. of water. When this systems cools by 30 F degrees it provides 120,000 Btu of thermal energy.

DISCUSSION

The pair of tanks absorbs radiation, stores energy, and distributes energy. The housing top serves as seat or couch. The upper structure serves as reflector, shutter, insulator, etc. Incidentally, the system reduces excessive heating and glare in the room, and, in summer, can be used to help keep the room cool. Some preheating of domestic hot water is provided and some back-up heating capability is included. The system is extremely simple. No antifreeze is used. The system operates even during periods of electric power failure.

However, it takes up some valuable space in the room and requires manual adjustment two or more times a day. Also the tilt of the reflectors, if correct for a certain midwinter month, is not quite correct for other months.

SYSTEM THAT INCLUDES INDOOR AND OUTDOOR, HINGED, REFLECTING-AND-INSULATING PLATES AND A GROUP OF OVERHEAD WATER-FILLED TANKS

PROPOSED SCHEME

Figure 1 shows the general idea, which is fairly similar to my Schemes S–166 and S–167 of 7/7/77 and somewhat similar to many earlier schemes such as S–10 of 3/22/73 and C–25 of 5/7/73. Also it is somewhat similar to Jeffrey M. Cohen's scheme described in US Patent 4,022,188.

A large curved reflector of cylindrical parabolic shape is used. It consists of two separate parts: outdoor part and indoor part. Both are curved and concave upward. Both have thick insulating backings and are slightly flexible so that they can be flattened for use at night as insulating shutters.

The outdoor part is mounted close to the south wall of the house and close to the ground. It is hinged, the hinge being close to the sill of the big south window. During the six hours in the middle of a sunny day in midwinter this reflector directs solar radiation slantwise upward, through the window and toward the storage system.

The indoor part likewise is hinged near the sill of the window. This part also directs radiation upward toward the storage system. Note that there is a clear space above the upper edge of this reflector, to admit some solar radiation deep into the room and to allow view of the outdoors.

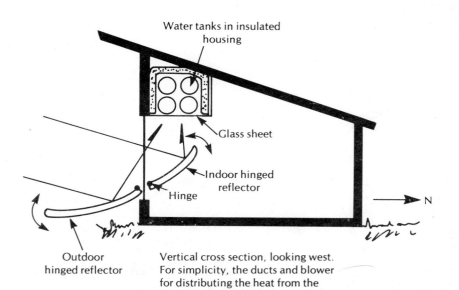

Water tanks in insulated housing

Glass sheet

Indoor hinged reflector

Hinge

N

Outdoor hinged reflector

Vertical cross section, looking west. For simplicity, the ducts and blower for distributing the heat from the tank–filled housing have been omitted.

The storage system employs some long horizontal cylindrical tanks filled with water. They are close above head-height, near the south wall. They contain, in all, several tons of water. They are inside an insulating enclosure that is generously sized to permit circulation of air between the tanks. A small blower can circulate room air through this space. The base of the enclosure is glazed with glass. The tanks contain no antifreeze; should the storage system ever cool below 40°F, a tiny electric heater within the insulating enclosure comes on automatically to prevent further drop in temperature.

The same system serves also the domestic hot water supply via a small heat exchanger incorporated in the storage system.

On cold nights the two reflectors are swung up, close to the window, to insulate it. During sunny days, the reflectors are oriented so as to reflect radiation toward the storage system, and, from week to week, slightly different tilts are used to accommodate the changing altitude of the sun at noon.

In the six warmest months, the indoor reflector is not needed and can be removed and stored in the garage. In the three warmest months the outdoor reflector also may be removed—unless the occupant wishes to use it to exclude radiation on the hottest days.

(Note: It is not necessary that the reflectors be flexible, i.e., for flattening against the window at night. The reflectors can be rigid and always curved if vertical adapter strips, or filler strips, are affixed at the E and W ends of the window to close the gaps there when the reflectors are swung upward against the window).

MODIFICATIONS

Scheme S-168b

Here the upward traveling radiation is absorbed by a conventional flat-plate panel and the hot water in this panel is circulated to a conventional water-type storage system.

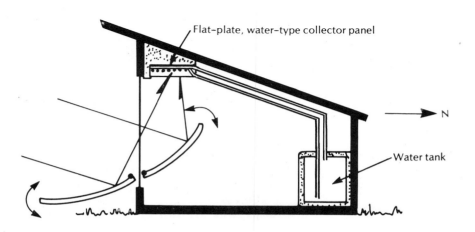

Flat-plate, water-type collector panel

N

Water tank

This scheme provides more storage and permits greatly reducing the size of the assembly near the top of the window. However, it is slightly more complicated.

Scheme S-168c

Here the absorber is cooled by air. Absorber and air are within a plenum, near top of window, that has a transparent plastic "floor". A small blower circulates the hot air from this plenum to a bin-of-stones. As before, the curved, thick, insulating reflectors can be swung so as to insulate the window at night or, in summer, swung so as to exclude solar radiation.

Notice that there are no pipes to leak, no water to freeze.

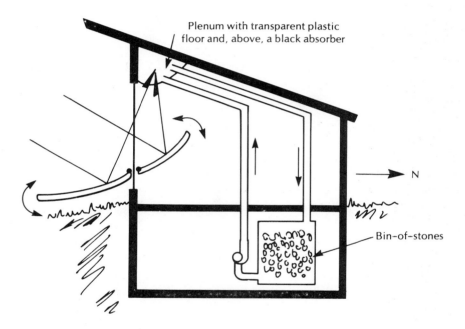

Plenum with transparent plastic floor and, above, a black absorber

N

Bin-of-stones

PASSIVE SYSTEM THAT EMPLOYS UPWARD-TRAVELING RADIATION AND TOTAL INTERNAL REFLECTION WITHIN TALL, WATER-FILLED GLASS TANKS AND CAN ACHIEVE VERY HIGH TEMPERATURE WITH VERY SMALL LOSSES

PROPOSED SCHEME

A long cylindrical parabolic reflector with east-west orientation is used to concentrate solar radiation and direct it upward toward the bottoms of a row of vertical cylindrical tall storage tanks of glass (or transparent plastic that can stand high temperature). An insulating housing encloses the tanks.

The radiation that is reflected upward toward the bottom of a tank has a large spread of directions. But as the radiation enters the tank, via its glass bottom, the spread of directions is reduced about 30% because of the excess of refractive index of glass (1.5) and water (1.3) relative to that of air (1.0). Most of the rays that are traveling upward within the tank and strike a side wall of the tank are totally internally reflected and continue onward in a generally upward direction. Thus, most of the radiation reaches the uppermost portions of the tanks and is absorbed by black vanes and black coatings here; accordingly, the water here is heated to very high temperature. Almost no heat can escape through the sides or tops of the tanks because of the excellent insulation used. Almost no heat can escape downward—even by radiation process—because it is intercepted by the lower-lying regions of water.

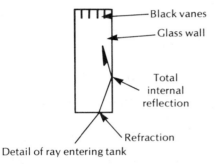

Detail of ray entering tank

- Black vanes
- Glass wall
- Total internal reflection
- Refraction

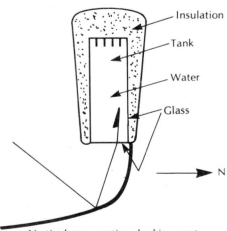

Vertical cross section, looking west

- Insulation
- Tank
- Water
- Glass
- N

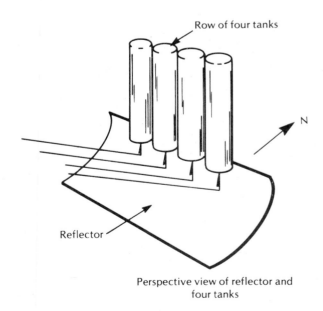

Perspective view of reflector and four tanks

- Row of four tanks
- N
- Reflector

In principle, the tanks could be designed so that the uppermost regions would reach, say, 500°F, if the water there could be prevented from boiling. In practice, temperature in the neighborhood of 200°F should be achievable easily enough. If a liquid having much higher boiling point were used, much higher temperatures could be achieved.

DISCUSSION

Some favorable features of the scheme are:

Because the liquid is transparent to solar radiation, and because the bottoms and sides of the tanks likewise are transparent to this radiation, practically all of the radiation entering the bottoms of the tanks travels all the way to the uppermost regions of the tanks. Total internal reflection plays a major role here. (If the tank walls were of steel there would be no total internal reflection and the radiation would be absorbed at lower locations, reducing the extent of stratification.)

Everything conspires to retain heat at the top of the tank, despite the very high temperature reached here. Every mechanism of heat-loss is almost entirely defeated.

No tracking is needed. The cylindrical parabolic reflector does its job well throughout about the five central hours of the day. (The tilt of the reflector is adjusted manually every few weeks.)

The water in the uppermost regions of the tanks is so hot that it could be used to drive Rankine engines or run absorption-type cooling systems.

The only major construction problem is to provide a tank bottom that is transparent and watertight and also can withstand high hydrostatic pressure. A thick plate of glass or plastic could be used. Or one could use a thin plate that is supported by a network of high-tensile-strength wires. The water, of course, must be very transparent, i.e., very clean.

MODIFICATIONS

Scheme S-170a

Instead of using a row of cylindrical tanks, use one large rectangular tank. But the tank walls will bulge unless they are very strongly made or are braced. The tank bottom is transparent. The walls are of glass or of steel that is faced with reflective foil.

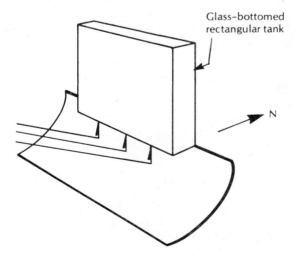

Glass-bottomed rectangular tank

N

AUGMENTED-HORIZONTAL-PIPE SOLAR COLLECTOR CONSISTING OF A ROW OF CYLINDRICAL PARABOLIC CLIP-ON REFLECTORS MOUNTED ALONG A FOUR-PURPOSE EAST-WEST BLACK STEEL PIPE

Scheme C-92
7/7/78

SUMMARY

The heart of the collector is a long, straight, horizontal, east-west, sturdy, black steel pipe that is put to **four** uses: it (1) absorbs radiation, (2) carries coolant (water), (3) supports a row of small cylindrical parabolic reflectors, and (4) controls their tilt. The builder installs such a pipe, then obtains (e.g., by US mail or United Parcel Service) a set of such reflectors and clips them onto the pipe securely. He installs, in all, several such pipes, each augmented with reflectors. He fills the pipes with water, starts the circulation pump, and the system is running. Every week or two he operates levers that turn the pipes a few degrees about their long axes so that the reflectors' tilt will be optimum for the coming week or two.

The system is very cheap—the cost of the equipment is very low and the cost of installing it is very low.

The many advantages of the system are listed. It may be near-ideal for retrofit.

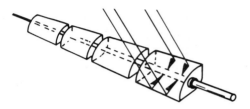

INTRODUCTION AND ACKNOWLEDGMENT

Most coolant-carrying pipes that run from collector to storage system are well insulated, yet lose significant amounts of heat. In mid-1978 I learned from S. C. Baer the idea of "beefing up" such pipes so that they will be energy gainers, not losers. Here I try to carry his idea further. I postulate a system that has no panels—just the pipes and several small reflectors clipped to them. (See also my 6/8/78 letter to S. C. Baer, describing Scheme C-90.)

Could it be that designers of solar collection systems have started the wrong way around? They started with big black sheets of metal and then attached metal tubing to them; or they started with big shallow boxes and attached inlet and outlet air ducts. Or they started with a long, wide reflector and then installed a water-filled tube along the focal line.

In each case they started with a long wide object.

Could it be that they should have started with a pipe, then thought what kinds of equipment to attach to it?

PROPOSED SCHEME

Here we start the other way around; we start with a long straight sturdy pipe and put it to **four** uses. Here is one possible design.

Install a 32–ft.–long, 1½–inch–dia., black, galvanized steel pipe along a near–horizontal east–west line just below the first story windows of the house in question. Slope the pipe ½–deg. downward toward one end, so that water can be completely drained from it at the end of a sunny day. Hold the pipe 8 inches from the south vertical wall of the house by means of eye–bolts or brackets 8 or 12 ft. apart on centers. The pipe passes through the "eyes" and is left free to rotate about its axis. At each end of the pipe there is a water connection consisting of a rubber coupling; it is flexible enough to permit the pipe to be rotated about 45°. A one–foot–long control arm is firmly attached to the west end of the pipe and another is attached to the east end. By turning these arms the homeowner can rotate the pipe a few degrees and thus change the tilt of the set of reflectors a few degrees.

After the pipe is in place, procure (from, say, some solar mail–order company, via the US mail or via United Parcel Service) a set of seven reflector assemblies designed to be clipped securely to such a pipe. Mount them on the pipe, bring them all to the same tilt, and tighten the various set screws (locking screws).

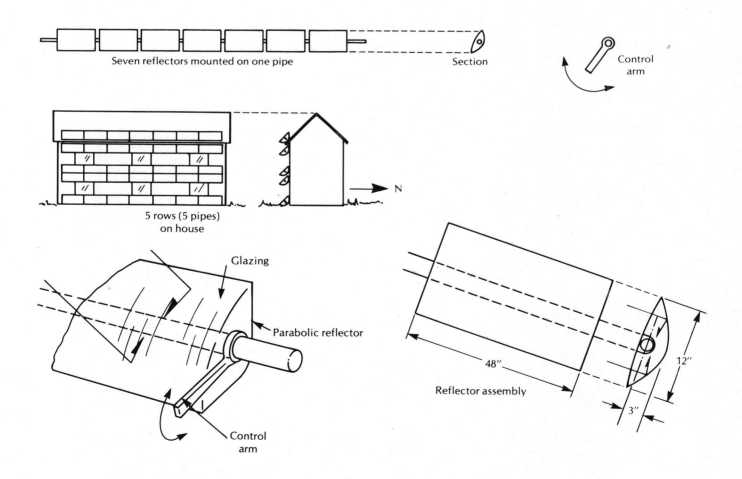

Seven reflectors mounted on one pipe   Section

Control arm

5 rows (5 pipes) on house

N

Glazing

Parabolic reflector

Control arm

Reflector assembly

48″

12″

3″

Each reflector assembly is 4 ft. long, 1 ft. wide, and about 4½ inches deep. The heart of each assembly is a piece of 80-to-85% reflective aluminized sheet bent to cylindrical paraboloidal shape with 6-inch radius of curvature at the center and 3-inch focal length. The sheet has a stiffening, strengthening, insulating backing (e.g., sprayed-on or glued-on pad of plastic foam). The two end pieces, which are reflective, are notched and can embrace the pipe and can be affixed to it by means of set screws. Around the four main edges of the sheet, there is a lightweight, somewhat stiff, frame of aluminum or plastic.

After the reflector assemblies have been mounted on the pipe and secured to it, all at the same tilt, a glazing sheet is clipped onto each assembly, covering its aperture and thus excluding rain, snow, leaves, etc., and trapping warm air. The glazing is of tough plastic. It clips on, and can be replaced every few years, if necessary.

The tilt of the set of reflectors is now adjusted (by means of the above-mentioned control arms affixed to the ends of the pipe) so that the tilt is optimum with respect to the sun at 1½ hours before solar noon and 1½ hours after solar noon; that is, it is optimum when the elevation angle of the sun is about 4 deg. less than the noon angle. Accordingly the tilt is "off" 4 deg. at solar noon and "off" the other way by about 8 deg. at 2½ hours before or after solar noon. It turns out that, if the reflector is perfectly made and is tilted as specified here, about 99% of the radiation specularly reflected by the reflector in the five-hour midday period will reach the pipe. Throughout an additional 1 or 2 hour period a moderate amount of solar energy reaches the pipe. At other times most of the reflected rays miss the pipe and accordingly the pipe is drained automatically: it is empty until 9:00 a.m. and after 4:00 p.m. and also at any other time when freeze-up might occur.

On a typical two-story house with east-west main axis, five such linear arrays may be used: one below the first story windows, two between first and second story windows, one above second story windows, and one along the lower edge of the roof. If each pipe is 32 ft. long and each supports seven 4-ft. reflectors, there are $5 \times 7 = 35$ reflectors with a total area of $4 \times 35 = 140$ sq. ft.

OPERATION

Water is circulated through the pipes and to the storage tank or room radiators during that portion of the 9 a.m. to 4 p.m. day when the pipes are hotter than the storage tank. At other times the pipes are empty.

Every week or two each pipe is rotated slightly so as to match the tilts of the reflectors to the changing height of the sun at midday. Exception: no adjustment is needed in the 60-day period from Nov. 21 to Jan. 21 or from May 21 to July 21; in such periods the height of the sun at noon changes through a total angle of only 3°. All of the adjustments are made by a person standing on the ground; the very-high-up control arms are equipped with extension control rods that extend down almost to ground level.

In summer the equipment may be kept in use to heat domestic hot water. Or the reflector assemblies may be removed and stored in the basement.

BAD FEATURES

The reflector tilts must be adjusted every week or two (except near the solstices).

Because the reflectors are of focusing type, their collection efficiency is especially low on overcast days.

GOOD FEATURES

The pipes can be purchased locally and installed by a local plumber.

The basic layout and basic installation (of the pipes) can be done before the reflectors arrive on the scene. All of the hydraulic testing, too, can be done before the reflectors arrive.

The reflectors are so small and light that they can be handled easily and can be shipped by US mail or by United Parcel Service. They can be transported even in a small car.

They can be clipped onto the pipes by the homeowner. Each can be installed in 5 minutes.

They can be removed and stored in summer.

No one has to climb onto the roof. (the system *could,* of course, be roof mounted.)

Tilt adjustments are made from the ground.

If tilt readjustment is forgotten, no damage is done.

The main portions of the pipes need no insulation.

Draindown is simple, quick, foolproof.

No antifreeze is needed.

No heat exchanger is needed.

The system can be used directly for heating space or heating domestic hot water.

On sunny days collection efficiency is high even if the storage tank temperature is as high as, say, 160°F.

Because the reflectors are small and simple, they could be made by true mass-production methods—at the rate of one every 10 seconds, say.

The system may be near ideal for retrofit applications.

COST

My guess is that the reflector assemblies would cost, installed, $4 per sq. ft. and the entire collection system would cost $9 per sq. ft. of reflector area.

(A conventional water–type collector, installed, costs about $15 to $30 per sq. ft. of collecting surface, I guess.)

POSSIBLE VARIATIONS

Install the augmented pipes along fences, garages, etc.

Employ pipes that have elliptical, not round, cross section. Orient the ellipse (of 2–to–1 shape, say) so that its long axis aims toward the sun. This geometry increases the capture of imperfectly aimed rays reflected by reflective regions near the two long edges of the reflector. Thus, the system performs well for a few more minutes each day. Also, the chance that the reflectors will slip rotationally on the pipes is eliminated, because the pipe is no longer round.

Reduce radiation losses by applying a *selective* black coating to the pipes. For example, wrap the pipes with the 6–inch–wide, adhesive–backed, selective–black "tapes" sold by MPD Technology Corp. of Waldwick, New Jersey.

Modify the designs and sizes of the reflectors so that they can be nested together, for shipment, in groups of four. Or go to the other extreme: assemble each reflector completely, including glazing, at the factory. Then, at the house site, install the reflectors on the pipe by disconnecting one end of the pipe and sliding the reflectors over this end.

Rotate the reflectors, not the pipe. Design the reflectors so that their ends interlock, so that turning one reflector turns many. (Suggested by R. T. Kriebel.) Affix to the pipe a set of indent plates, or sleeves, that precisely determine the various reflector tilts successively invoked throughout the year.

Make a twice–size version for industrial use.

Design, for use with system employing air as coolant, a corresponding system with 6–inch–dia. pipe and large (6 ft. × 2–ft.) reflectors. Cause air to flow within the pipe.

COLLECTOR EMPLOYING AN ARRAY OF CYLINDRICALLY FOCUSING DEVICES, THE ARRAY HAVING NO CONVENTIONAL TRACKING SYSTEM BUT BEING MOUNTED ON A FLOATING PLATFORM WHICH IS SLOWLY ROTATED

Scheme C-58
5/11/76

SUMMARY

Figure 1 shows the proposed scheme. The tilted array is mounted on a float which is rotated slowly (one half revolution per 12 hours) to provide the appropriate tracking. Because the individual devices are *not* individually rotating, many complications are avoided and the boxes may be packed tightly together to save space.

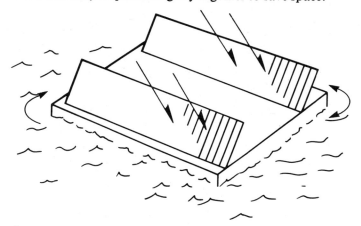

Perspective view of float with two tilted arrays of collector devices

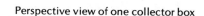

Perspective view of one collector box

INTRODUCTION

Suppose one has 100 of the Northrup collector boxes, each of which employs a long cylindrical fresnel lens, with a coolant-carrying pipe along the focal line. Normally, these devices are spaced a few feet apart and are oriented with the long axes parallel to the earth's axis. Each box is turned individually, by a conventional tracking mechanism, so as to always face the sun. (Why the space between boxes? To allow room for each to turn individually and to prevent tendency (near dusk, say) for one box to shade a neighbor.) The liquid in the pipes carries energy, at, say, 200° or 300°F, to a tank.

PROPOSED SCHEME

Mount the boxes on a float that is moored in a pond (in Mass., say) protected from high winds. Pack the boxes close together, to save space. Provide *no* conventional tracking system. Instead, turn the whole float slowly—at the rate of one half revolution per 12 hours to keep each box aiming at the sun. If one wishes to, one may make gross changes in tilt of boxes a few times a year; tilt-angle from horizontal might be set at 70° in December, 35° in June. But a fixed, 45° tilt would work pretty well throughout the year. (A system used near the Equator could be set horizontal permanently; wind problems would be much less.) Suitable "ponds" (6 in. deep, say) could be constructed nearly anywhere, even on the roofs of big horizontal-roofed buildings.

Note I have recently learned that, for several years, C. B. Cluff of the University of Arizona has worked on floating, rotating arrays of concentrating collectors. His recent work is described in *Proceedings of the Am. ISES August 1978 Conference* in Denver Colorado, Vol. 2.1, p. 929.

POOR MAN'S CYLINDRICAL-LENS-TYPE COLLECTOR FOR REGIONS NEAR THE EQUATOR: SYSTEM EMPLOYING A WATER LENS

Scheme C–57
5/10/76

SUMMARY

Using a fairly taut transparent plastic sheet draped over parallel horizontal bars, it is easy to make pools of water that will focus solar radiation onto a set of parallel, black, coolant-filled pipes some distance below. Such a (cheap) collector—reminiscent of the successful but costly Northrup collector—should work well near the equator, where the sun so often is nearly overhead.

INTRODUCTION

The Northrup collector box, with cylindrical fresnel lens, is well known and highly successful. But it is expensive, especially as the box is usually oriented with the long axis parallel to the earth's axis and tracking is required.

A simpler way of using the Northrup collector might be especially successful in regions near the equator. If the box axes are E–W, and box aims roughly straight up, it performs excellently (in March and September) with no tracking at all. Consequently, such boxes can be ganged together, say, in groups of five; the coolant pipes, too, can be assembled at the factory in gangs. Every few weeks the tilt of the gang is manually readjusted slightly. Even on Dec. 21, say, the collector works fairly well inasmuch as the sun stays within a few degrees of the same plane throughout the central five hours of the day. (See my 5/10/76 letter to L. Northrup.)

Can one devise, for use near the equator, an even simpler system? I think so.

PROPOSED SCHEME

Keep the gang of lenses horizontal at all times, with axes E–W. When adjustment is needed, shift the gang of pipes N or S slightly, as required to keep the pipes along the focal lines. Replace the fresnel lens with a water-and-plastic-sheet lens, formed by draping a slightly taut sheet of durable, transparent, waterproof plastic over a set of parallel, horizontal, uniformly spaced, E–W bars (about 1 ft. apart on centers), and flooding the sheet with water. The weight of the water makes the sheet sag (1 or 2 in., say) between bars. Lens-like pools of water are formed, and produce line-like images (1 or 2 ft. below). A set of black tubes, carrying coolant, lie along these images, absorbing the radiant energy. The position of the gang of pipes is adjusted slightly every few weeks. If the plastic sheet is kept taut by means of a heavy hanging bar attached to the non-fixed edge, the tension in the sheet remains constant, and so does the focal length, irrespective of the temperature–and–moisture history of the sheet. Rippling of the water surface may be reduced by wind baffles, by an oil film, or by laying a sheet of plastic (Teflon, say) on the water. An oil film will reduce evaporation. Rain can make up for evaporation losses. Notice that the lenses, besides focusing the direct solar radiation, exclude wind, rain, leaves, bugs.

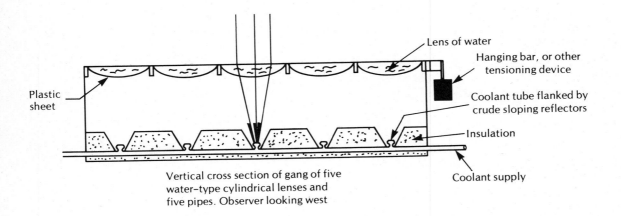

Plastic sheet

Lens of water

Hanging bar, or other tensioning device

Coolant tube flanked by crude sloping reflectors

Insulation

Coolant supply

Vertical cross section of gang of five water–type cylindrical lenses and five pipes. Observer looking west

Warning: Lens power is small (focal length large) because refractive index of water is much less than that of glass. For higher index, use clear paraffin oil, say.

COLLECTOR EMPLOYING A TRACKING SHEET-LIKE ARRAY OF RECTANGULAR, FRESNEL, SPHERICAL LENSES

Scheme C-50
4/3/74

SUMMARY

A large, sheet-like array of rectangular, fresnel, spherical lenses of molded glass (or plastic) focuses the sun's rays onto a corresponding set of tiny spots on a set of black, liquid-filled tubes and heats the liquid to 400°F (guess). A simple tracking system is provided. The sheet-like array of lenses not only focuses the sun's rays onto the pertinent spots on the tubes but also (a) serves as a window to exclude rain and wind and reduce heat-loss by convection, and (b) provides greenhouse-effect trapping of re-radiated energy.

Most of the radiation from the blue sky passes between the tubes and enters the room, illuminating it and helping heat it.

PROPOSED SCHEME

The accompanying figures show the main features of the design.

Figures 1 and 2 show a single rectangular, fresnel, spherical, glass (or plastic) converging lens that is about 2½ in. × 1 in. in area, with the short axis horizontal. Because the lens is of fresnel type (having, in all, three facets), its thickness is modest (about ¼ in.). When the lens is placed normal to the sun's rays, it produces an image (of the sun) that is 6 in. from the lens and is about 1/16 in. in diameter.

Figure 3 shows a linear sequence of such lenses, molded as a unit, and Fig. 4 shows a two-dimensional array of such lenses, molded as a unit.

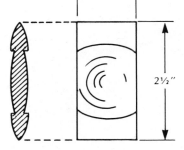

Fig. 1. Rectangular fresnel spherical lens. Two views

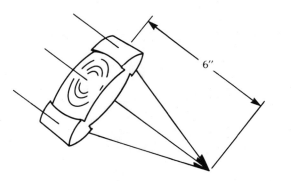

Fig. 2. Perspective view of lens

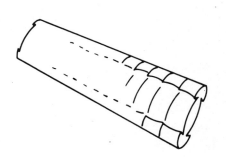

Fig. 3. Linear sequence of lenses, molded as a single unit

Figure 5 shows such a planar array and a drive system for moving the array parallel to itself. The system employs guides, actuator, bar, and cams. During the daylight hours the drive system moves the array gradually westward and upward, then westward and horizontally, then westward and downward—in such manner that the solar images remain nearly stationary. Slight readjustment is needed weekly—slight changes in the vertical positions of the cams.

Figure 6 shows a liquid-filled tube 0.3 in. in ID. The tube is of metal and has a 0.3-inch-thick insulating jacket that is white in color. Every inch along the jacket (along the portion nearest the array of lenses) there is a hole 0.2 in. in diameter; the hole exposes a 0.2-inch-diameter portion of the tube; this portion, or spot, is coated with a selective black coating.

Figure 7 shows an array of such tubes. The tubes are horizontal and are 2½ in. apart on centers. The exposed portions of the tubes comprise a network with spacings of 1 in. horizontally and 2½ in. in direction perpendicular thereto.

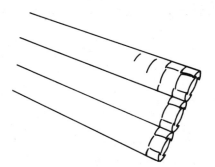

Fig. 4. Planar array of lenses, molded as a single unit

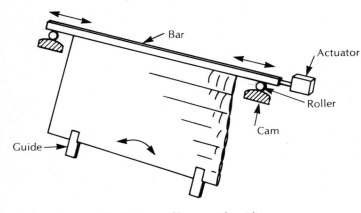

Fig. 5. Array of lenses with guiding and actuating system, to provide tracking throughout 7 hours of a sunny day

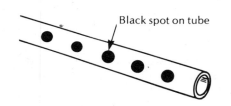

Fig. 6. Detailed view of segment of one tube. The insulation is not shown.

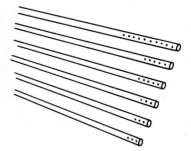

Fig. 7. An Array of many parallel water-filled tubes, with black absorber spots

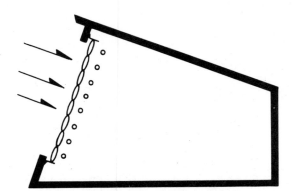

Fig. 8. Vertical cross section, looking west, of house with planar array of lenses focusing suns rays onto spots on adjacent array of horizontal tubes

Figure 7 shows the system applied to a small house. The figure is highly simplified. Manifolds, storage systems, etc., are omitted.

OPERATION

On a sunny day, each lens focuses direct solar radiation onto an exposed spot on a tube, and as the angle of the sun gradually changes the lens array is gradually translated, parallel to itself, so as to keep the focal spots approximately fixed in position. (The actuator can operate continuously or intermittently; one actuator can control a lens array having an area as large as 6 ft. × 10 ft., or even larger.)

The tubes receive energy, and the liquid in the tubes is heated and carries energy to a storage tank. I estimate that temperatures as high as 400°F can be reached.

The area of the individual lens is $(1 \times 2\frac{1}{2}) = 2\frac{1}{2}$ in.^{2}, and the area of an exposed portion of tube is 0.03 in.^{2}; thus, there is a geometrical compression of $(2.5/0.03) =$ about 80, in terms of area. The selective coating has an a/e ratio (ratio of absorptivity in solar band to emissivity in re-radiation band) of about 12. Thus collection efficiency is high, despite the very high temperature achieved. (Much of the heat that is lost from the tubes remains in the room, helping heat it.)

Notice that the array of lenses performs many functions:

1. The array concentrates the direct radiation, as discussed above. This is its most important function.

2. The array, comprising a single, large, air–tight sheet, prevents convection of air through the plane of the array. Also it prevents passage of rain, dust, insects. In short, it serves as a window (but without introducing any reflection losses other than those inherent in a simple lens).

3. Having high absorptivity for 4-to-40 μ m radiation, the array absorbs much radiation flowing from the exposed portions of the tubes or from other warm objects thereabouts. In short, it provides greenhouse–effect trapping.

Notice also that most of the radiation that comes from the blue sky does *not* strike the tubes: it passes between them and enters the room. Thus, the level of illumination in the room is about one fifth of the level that would exist if the entire collection system were replaced by a simple window.

The focusing effect of the lenses is greatly degraded when the incident rays are at angles exceeding about 35 degrees from the normal. Accordingly, collection efficiency is poor at times more than about 2½ hours from noon.

The array of lenses would be fabulously expensive unless a very large sum were invested in mass–production machinery.

MODIFICATIONS

Scheme C-50a

This is the same as above except that a manual override on the vertical position of the array of lenses is provided. The house occupant may at any time pull on a rope that raises this array 1 inch higher than would otherwise be the case. The consequence is that the focused rays, instead of striking the tubes, pass between them and enter the room. Thus, the room may be warmed quickly on a sunny day.

Scheme C-50b

Here cylindrical lenses are used instead of spherical lenses, and a full-length linear absorption strip is exposed along each water-filled tube. When such an optical scheme is used, the exact east-west position of the array of lenses becomes irrelevant. Alignment and tracking become somewhat simpler. Manufacturing suitable fresnel lenses becomes simpler. (One could use the 10 ft. × 1-ft. lenses already being mass-produced by Northrup, Inc., of Hutchins, Texas.) However, the total emitting areas of the tubes would be greater and consequently heat-loss from the tubes would be greater.

Scheme C-50c

Here the array of lenses always remains stationary and the array of tubes is moved so as to provide tracking. Using this scheme, the designer finds it easier to provide tight seals around the edges of the array of lenses.

Scheme C-50d

Here a thick, insulating, roll-down curtain is installed just north of the array of tubes. When this curtain is pulled down it defines, in conjunction with the array of lenses, a plenum in which stray heat from the array of tubes accumulates. A forced stream of air can carry this heat to a bin-of-stones in the basement. At night the same curtain may be used to reduce the amount of heat that leaks from room to outdoors via the array of lenses.

SYSTEM EMPLOYING AN OUTDOOR, BELOW-WINDOW-LEVEL, CONCAVE REFLECTOR AND A CANOPY-ENCLOSED WATER-TYPE BLACK ABSORBER

Scheme S–10
3/22/73

PROPOSED SCHEME

Outdoors, near the lower part of the vertical south wall, there is a 28-ft.-long, horizontal, cylindrical, aluminum-faced reflector that directs the sun's rays upward toward a water-type black absorber that is well insulated at top and sides by an enclosing canopy that includes much urethane foam and is insulated on the underside by two spaced sheets of glass. Glass, rather than plastic, is used in order that all of the 4-to-40-micron radiation emitted in downward direction by the black absorber will be intercepted (trapped). The hot water from the collector is circulated by a small centrifugal pump to an insulated steel tank in the basement. Heat from this tank is distributed, when needed, to the rooms by any conventional means. The cylindrical reflector is hinged along its upper edge and the tilt of the reflector is changed manually from month to month.

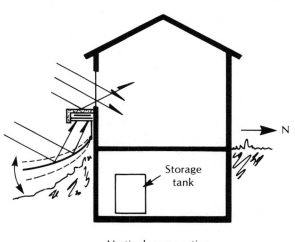

Vertical cross section, looking west

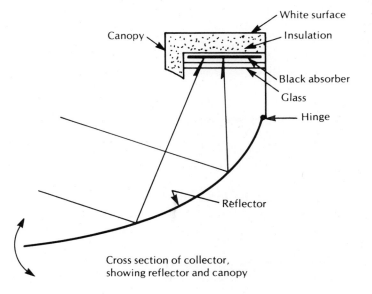

Cross section of collector, showing reflector and canopy

Some good features of the design are:

The collector provides 3-to-1 concentration.

No tracking is required, but the reflector tilt must be adjusted manually every few weeks if high collection efficiency is to be maintained.

The collector is readily accessible.

It is situated low enough down so that the south windows are free to receive much direct radiation and occupants of the room can enjoy a view to the south.

Heat-losses from the black absorber are very small. The canopy blocks upward heat-flow, and, below the absorber, the two spaced glass sheets, aided by beneficial thermal stratification in the trapped air, block downward heat-flow.

The sloping reflector sheds rain and leaves, and the upper part of it, at least, sheds snow.

As regards danger of freeze-up, the black absorber could consist of an extruded mat of synthetic EPDM rubber, such as the Bio-Energy Systems, Inc., mat which includes integral tubes 0.7 inch apart on centers. Freezing does not affect this mat. Outdoor headers are not required; the rubber tubes can extend into the basement, and the headers can be located there. Alternatively, a conventional absorber could be used and a small electrical heating strip could be used to insure that the temperature of the absorber does not fall below 40°F. Or antifreeze could be used.

MODIFICATIONS

Provide several collectors on several stories of a multi-story building.

Make the collection system rigid and have the entire system swing about a hinge along the upper north edge. Adjust the orientation every few weeks. Such a system collects much energy even in summer, whereas the basic Scheme S-10, employing a canopy that remains in fixed orientation, collects little radiation in summer (because the aperture between reflector and canopy is smaller then).

Place the (large, steel-bottomed) storage tank in the attic. Arrange for the steel bottom to serve as ceiling for the central hallway of the first story. Provide pipes between collector and storage tank, so that hot coolant will circulate to the tank passively (by gravity convection). Then the entire system is passive and operates without use of electrical power. Heat flows from the storage tank to the hallway etc. by radiation. Such system may be ideal for a weekend cottage at a remote site in Vermont—the solar heating system performs routinely even when the owner is away, the cottage is empty, and the supply of electrical power fails.

Other modifications are described in the reports S-10 to S-40a and C-10 to C-29, written early in 1973.

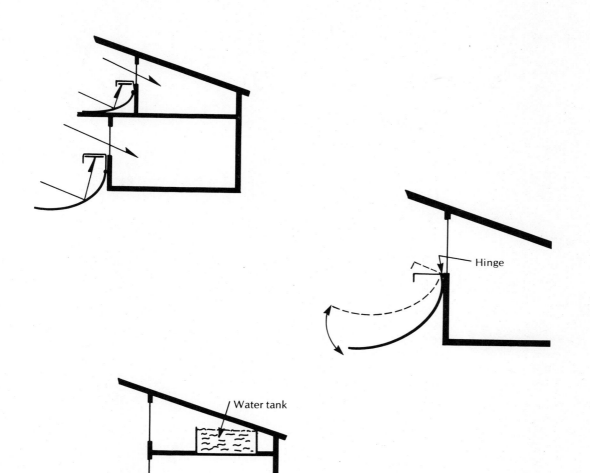

Water tank

Hinge

Related Inventions by Others

A somewhat similar collection system has been invented by J. M. Cohen of 2014 Locust St., Philadelphia, PA 19103. He patented his invention; see US Patent 4,022,188 of 5/10/77. A related scheme was patented by D. J. Lightfoot; see US Patent 4,003,366. Others also have described collectors of somewhat similar design.

PART 5

Storage Systems Employing Conventional Materials

INTRODUCTION

Here I discuss storage systems that employ conventional heat-storage materials such as stones, water, etc. Phase-change materials are discussed in the following Part.

Some of the pertinent goals here are: increasing the thermal capacity of the storage system, reducing the resistance to flow of coolant and thus reducing the amount of power needed to circulate it, avoiding unnecessary reduction in thermal stratification, increasing the rate of heat input to (or output from) a tank of water, and making such tank itself serve as a heat exchanger.

Bins-of-stones are discussed first, then water-filled tanks.

BIN-OF-STONES EMPLOYING THREE SIZES OF STONES IN THREE ZONES TO INCREASE THERMAL CAPACITY AND PNEUMATIC CONDUCTANCE

Scheme S–104
9/2/76

PROPOSED SCHEME

Divide the bin-of-stones into three zones and use different size stones in each. In the top, middle, and bottom zones, use stones having average diameters of 1 in., 3 in., and 5 in., respectively. The hot air from the collector is blower-driven *downward* through the bin; thus, passing first through the quantity of small stones and last through the large stones. Thermal performance is improved and the requirement on blower power is reduced.

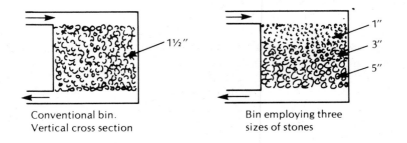

Conventional bin.
Vertical cross section

Bin employing three
sizes of stones

DISCUSSION

Practically all designers of bins-of-stones call for use of just one size of stones. This is true, for example, of the designers whose excellent articles on bins-of-stones appear in *Solar Age*, April 1978, p. 23, 24, 25, 40. Yet a little thought shows that from the standpoint of performance the use of a single size of stones is unwise, especially if the bin is very large. If one mentally divides a large bin into three horizontal zones and considers the functions and penalties associated with each, one soon realizes that:

The top zone sees most of the action—it plays the most important role. In midwinter, when the amount of energy in the bin is small, the solar energy collected in a period of bright sunshine is delivered to the top zone of the bin, and when the rooms next need heat, heat is extracted from this top zone. Again and again it is the top zone that is active.

The bottom zone is seldom active. Usually it takes up energy only when the two superior zones are already fairly full of energy, which occurs mainly in fall and spring. The bottom zone may be regarded as a kind of stand-by component, storing energy only when there is near-surplus.

Yet the pneumatic conductance of the bottom zone is no greater than that of the top zone.

In summary, the bottom zone is much less useful, yet imposes the *same* burden on the blower.

Clearly the bottom zone should be filled with stones that are larger than the stones of the top zone. This only slightly reduces the rate of heat uptake by the bottom zone during crucial times of winter but considerably increases the pneumatic conductance of the bin as a whole. If, in a standard bin, all of the stones are 1½ in. in diameter, a modified bin should contain, in the top zone, stones of slightly smaller diameter (to provide significant increase in heat-exchange surface area here) and, in the middle and bottom zones, stones of progressively greater diameter (to increase the conductance here). The overall effect is improved uptake (or distribution) of heat and reduced amount of blower power needed.

Is the improvement worthwhile? Friends tell me that, ordinarily, the pneumatic conductance of an air-type collector is smaller than that of a bin-of-stones; accordingly, any increase in conductance of the bin-of-stones is of modest value. Perhaps what is really needed is to increase the conductance of both collector and bin-of-stones.

HOW TO DOUBLE THE THERMAL CAPACITY OF A BIN-OF-STONES WITHOUT INCREASING THE SIZE: REPLACE HALF THE STONES WITH AN EQUAL VOLUME OF WATER IN TANKS

Scheme S-102
9/2/76

SUMMARY

One can double the thermal capacity of a bin-of-stones storage system of an air-type solar-heated house—without increasing the bin size—by (a) replacing the stones in the bottom half of the bin with an array of small tanks of water, and (b) employing, whenever the stones are adequately hot, a small, steadily running blower to circulate air within the bin and equalize the temperature of stones and water. Pneumatic conductance is unchanged. Heat-transfer area is reduced, but not significantly. One enjoys the best of both worlds: the high thermal capacity of water and the large heat-transfer area of stones.

INTRODUCTION

Consider a 10-ft. × 10-ft. × 10-ft. bin that is filled with 2-in.-dia. stones. During a sunny day a blower drives hot air from the solar collector downward through the bin. When the rooms need heat a blower drives room air upward through the bin. The heat-exchange area of the system is large. The pneumatic conductance is adequate. Beneficial stratification of temperature within the bin occurs automatically. But the amount of heat stored is small, being only enough to keep the house warm for, say, one sunless January day. (In practice, many bins are much smaller and may provide only half-day carrythrough.)

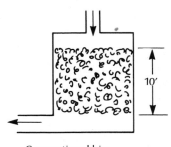

Conventional bin.
Vertical cross section.

METHOD OF DOUBLING THE THERMAL CAPACITY

Replace the stones in the bottom half of the bin with a compact array of water-filled, 55-gal. steel tanks. This doubles the thermal capacity of the system as a whole. If 1-in. spaces are left between tanks, the pneumatic conductance of the system remains high—it may even be higher than before.

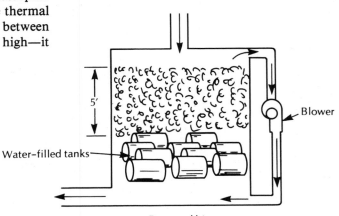

Water-filled tanks

Blower

Proposed bin.

The heat-transfer area of the system as a whole is reduced by a factor of about 2, inasmuch as the quantity of stones is halved and the surface area of the tanks is very small compared to that of the stones. Does this reduction hurt the performance seriously? Ordinarily not. During typical midwinter conditions the amount of heat in the bin is small and most of the action occurs in the upper part of the bin: it is the upper part that accepts the heat-load from the collector during sunny days and imparts heat to the rooms during cold nights. In midwinter the lower part of the bin seldom comes into play; it comes into play only when the bin contains much energy and can keep the house warm for a long time, and when long time-constants of heat transfer are acceptable. Note that the spontaneous, gravity-induced circulation of water within each tank helps the heat-transfer process.

Let us make one additional change: Let us provide a small blower that, ordinarily, runs steadily 24 hours a day circulating air vertically within the bin to equalize the temperatures of stone and water. This makes the water carry a bigger share of the energy storage and makes the stones carry a smaller share. The total amount of energy stored tends to be much greater. It may pay to have the blower shut off automatically when the stone cool down below 120°F, say, and on other occasions when sacrifice of stratification is highly undesirable.

Installation of dampers to prevent wrong-way, or bypassing, airflow may be necessary.

MODIFICATIONS

Scheme S-102a

Install the tanks just above (not below) the stones. This reduces the amount of beneficial stratification but has several advantages. Supporting the stones is made easier. The tanks are far more accessible—easier to inspect and service. When the entire system is quiescent, heat flows spontaneously, by gravity convection, from stones to tanks (but at the cost of reducing the stratification in the stones). A steadily running low-power blower may still be used, but it is far less necessary.

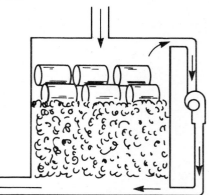

Bin with water-filled tanks at top

Scheme S–102b

Use just one tank—one big, large-area tank—and connect it, by pipes, to the domestic hot water system, so as to preheat this water directly. No heat exchanger is needed because the water in the tank is pure; no antifreeze is needed because the tank, being in an insulated bin in the basement, will never cool down to 32°F. A small blower is used to maintain a steady transfer of heat from the stones to the tank. The blower consumes only $10 worth of electric power per winter.

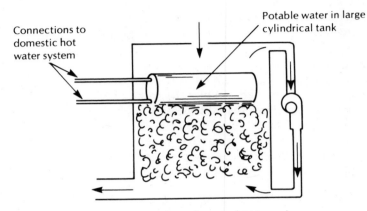

Bin with one large tank at top

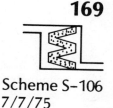

FOLDED BIN–OF–STONES THAT PROVIDES SHORT PATHLENGTH AND LARGE CROSS SECTION FOR AIRFLOW

Scheme S-106
7/7/75

SUMMARY

The proposed folded bin–of–stones provides a much shorter pathlength and much larger cross section of airflow. The pneumatic resistance is reduced by a factor of the order or 50; accordingly, the requirement on blower power also is reduced. The extent of thermal stratification achieved is much less, but the "percent-solar-heated" figure is increased.

PROPOSED SCHEME

The design principle used is to replace the conventional near-cubic bin–of–stones with a bin that is about 20 ft. wide and only about 1 ft. high, and then to fold this awkward-shaped bin so that, once again, the overall shape is near-cubic. A near-cubic shape is desired because the area of insulation then needed is small and the amount of basement floorspace preempted is small.

The accompanying sketches show the various stages in design change.

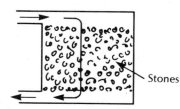

Stones

Conventional design,
vertical cross section

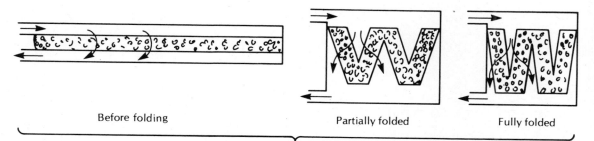

Before folding   Partially folded   Fully folded

Proposed bin—with short pathlength
and large cross sectional area in each case

Folding must not be so extreme that the V-shaped regions of air are eliminated. They are essential to shortening the pathlength in the stones and to increasing the cross-sectional area. Fortunately they take up little space. Also, they are easy to produce.

Whether the V-shaped regions of air run vertically or horizontally makes little difference. The accompanying sketches show the two alternatives.

In a typical installation employing six near-horizontal slab-shaped regions of stones, each slab is 10 ft. × 10 ft. × 1 ft. Thus the volume of each slab is 100 ft^3. and the overall volume of slabs is 600 ft^3. The mass is about 30 tons.

Near-horizontal wedge-shaped spaces for air input and air output are left between slabs. The spaces are maintained by layers of galvanized steel netting held apart by long, wooden, tapered stringers, such as trimmed trunks of sapling, or trimmed branches of trees. The spacers are 3 in. thick at one end and 1 in. thick at the other. Typical length: 9 ft. Spacers are 1 ft. apart on centers. The netting is secured all around the edges to a strong rigid structure, namely the main housing of the bin; thus, the netting is under tension and does not sag much; thus the spaces remain open. Steel structures could be used to provide the spaces, for example, channel irons, or bundles of angle irons, could be used; they would perform well, but they may be needlessly expensive.

The overall dimensions, including housing and insulation, are: 8 ft. high by 12 ft. long by 12 ft. wide; nominal pathlength provided: 1 ft.; effective cross-sectional area through which air flows: (6 slabs) (100 ft^2. per slab) = 600 ft^2.

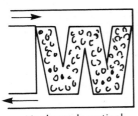

Air channels vertical

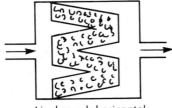

Air channels horizontal

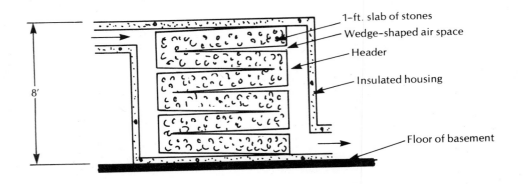

1-ft. slab of stones
Wedge-shaped air space
Header
Insulated housing
Floor of basement

8'

Vertical cross section of folded, short-pathlength bin-of-stones

COMPARISON WITH CONVENTIONAL BIN

The proposed bin is superior in several ways to a conventional bin containing the same mass of stones:

- The pathlength is less by a factor of about 6. The cross-sectional area of airflow is greater by a factor of about 6. Accordingly, the pneumatic conductance is greater by a factor of about 36, and the requirement on blower power is reduced.

- Because a smaller pressure of air is required to maintain the flow, the leakage of air from the bin or from the collector, ducts, etc., is reduced. Tolerances on air-tightness may be somewhat relaxed.

- Because less blower power is needed, a smaller and cheaper blower may be used, it takes up less space, is quieter, and uses less electric power. (But the improvement may be small, as explained in a following paragraph.)

- Because the pneumatic conductance of the bin itself is so very high, and because the bin is in the basement, beneath the rooms, much heat can be delivered to the rooms by gravity convective flow. That is, even if the electric supply fails and forced circulation of room air through the bin is halted, the rooms can, nevertheless, be kept at least fairly warm by natural circulation of room air through the bin. (A conventional bin has such low conductance that little air can circulate through it by gravity convection.)

- Because, with the pathlength in the stones shortened to 1 ft., the amount of thermal stratification achieved is small, the designer may arrange to have the cool air from the rooms circulate through the bin in the *same* direction that the hot air from the collector circulates. Under some circumstances this can simplify the overall operation of the system.

Note that the designer might decide, using the proposed scheme, to increase the pneumatic conductance only a little and to change to a smaller size of stones. Such change increases the heat-transfer area and improves collection efficiency and facilitates distribution of heat to the rooms. Yet, thanks to the folded design, the pneumatic conductance is still comfortably high.

TWO MAJOR LIMITATIONS OF THE PROPOSED SCHEME:

Professor E. J. Carnegie of California Polytechnic State University at San Luis Obispo has kindly pointed out to me that the scheme proposed here has two important limitations: (1) if the main pneumatic resistance is that of the collector and ducts, reducing the resistance of the bin-of-stones helps only a little, and (2) providing the special air spaces can add considerably to the materials and labor expenses of building and filling the bin.

MODIFICATION

Scheme S–106a

Same as above, except use a bin that has the shape of a hollow tapered cylinder. Here the increase in pneumatic conductance may amount to a factor of 10.

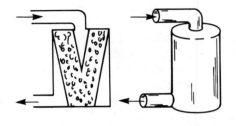

IMPROVING THE THERMAL PERFORMANCE OF A THICK INTERNAL MASONRY WALL BY PROVIDING VERTICAL CHANNELS IN THE WALL FACES

Scheme S–53
9/19/78

PROPOSED SCHEME

The thermal performance of thick masonry walls inside a solar heated house can be greatly improved by deeply grooving the walls. If the builder provides many deep vertical grooves, channels, or slots in such wall, the area of heat-transfer surface may be doubled and the average pathlength (within the wall) for heat input or output may be halved. Consequently, heat delivered by direct solar radiation or by warm air penetrates faster and deeper into the wall; even the innermost regions within the wall participate promptly in the storage process. The improvement applies not only to conduction and convection but also to radiation. The improvement applies mainly to heat input, but it applies to a moderate extent to the (more leisurely) process of heat output.

Using such channels, the designer may find it feasible to use even thicker walls and thus provide an even greater amount of prompt thermal storage.

The cost of providing vertical channels—when the wall is being built—is small.

The channel may be useful in additional ways. Electrical wires or water pipes may be concealed in the channels. Brackets for holding shelves or reading lamps may be installed in the channels.

Horizontal channels would be less useful because little gravity-convective airflow would occur. There would be no chimney effect.

One could provide vertical channels that are situated along the vertical midplane of the wall, entirely out of sight. Ports for air inlet and outlet would be provided. But the internal channels would be inaccessible. If a mouse were to die in such a channel, his death would long be remembered.

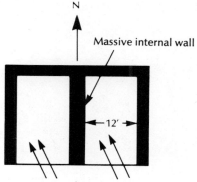

Horizontal cross section
of passively solar heated house
having massive walls

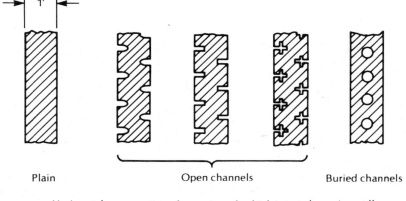

Plain   Open channels   Buried channels

Horizontal cross section of a portion of a thick internal massive wall

HOW TO COAX MORE HEAT FROM A NOT-VERY-HOT WATER-TYPE STORAGE SYSTEM BY PROVIDING A SMALL SPACE BETWEEN THE TANK AND THE ENCLOSING INSULATION AND CIRCULATING ROOM AIR THROUGH THIS SPACE

Scheme S-107
9/15/76

SUMMARY

Designers of water-type storage systems for solar heated houses seem to have overlooked one small asset: the large exterior surface of the tank proper. By leaving a slender space between tank proper and the enclosing insulation, and by arranging to blow air through this space when the tank is too cold to heat the house in the normal way via baseboard radiators, the designer can arrange to coax yet a little more heat from the tank. As a corollary, the tank will become even cooler and subsequent solar-energy collection will proceed more efficiently.

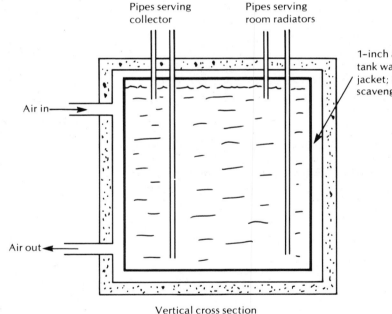

Pipes serving collector

Pipes serving room radiators

1-inch air space between tank wall and insulating jacket; airflow here can scavenge much heat.

Air in

Air out

Vertical cross section

INTRODUCTION

Here we consider a house the solar heating system of which includes a large, insulated, water-filled tank in the basement. We assume that, when the tank is very hot, water is circulated from it to the room radiators to heat the rooms, and when the tank is not very hot, it stands idle, unused, and the furnace is turned on, sending hot water to the radiators.

It would be helpful if the tank, when only moderately warm, could continue to supply *some* heat.

An elegant but expensive way to solve the problem is to provide two sets of baseboard radiators: one served by the storage tank and the other served by the furnace. The storage tank can then continue to supply a little heat even when it is only moderately warm and even when the furnace is running. Such a scheme is used in the Massachusetts Audubon Society Gift Shop in Lincoln, Mass.

The scheme described below is simple and may be cheaper.

PROPOSED SCHEME

Before applying the insulation to the big water-filled steel tank, wrap the tank with crinkled, or wavy, chicken wire; then apply the insulation. An airspace about 1 inch thick between tank wall and insulation remains. Install two small ducts (say, 4-inch-dia. cloth ducts with no insulation) that carry air to and from the 1-inch space, and to and from the front hallway or other central and communicating region. Install a small fan in one of the ducts and connect the fan to a timer that will keep the fan *on* from 8:00 p.m. until 8:00 a.m., so that throughout the night, a small amount of useful heat will be delivered to the living area of the house. The tank will become especially cool, and collection on the next sunny day will proceed at especially high efficiency.

MODIFICATIONS

Scheme S-107a

Here we apply the same general idea to a storage tank that is rectangular and of concrete. We provide a 1-inch space between tank proper and the enclosing insulation and we extract a small amount of heat from this space throughout the night.

Scheme S-107b

If the concrete tank has a liner, we provide, by use of wire netting and crude but strong spacers, a 1-inch space between the liner and the concrete walls and bottom of the tank. We extract heat from this space throughout the night.

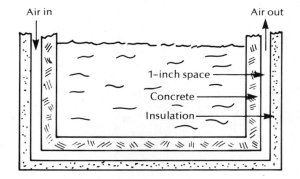

Airspace is between concrete and insulation

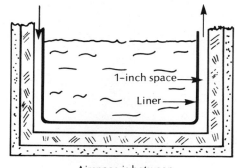

Airspace is between liner and concrete

Scheme S-107c

Here we proceed as in one of the above-described schemes, but in addition we make the space somewhat larger (say 3 inches thick), and we install a few hundred pounds of stones, or steel bars, in this space. Then the stones, or bars, gradually heat up throughout the day, when no heat is being extracted from this space, and they contribute much heat rapidly to the small airstream serving the front hallway at night.

SCHEME WHEREBY A STREAM OF HOT WATER BEING RETURNED TO A STORAGE TANK IN WHICH THE WATER IS THERMALLY STRATIFIED WILL AUTOMATICALLY JOIN THE TEMPERATURE-MATCHING STRATUM

Scheme S–50
8/5/75

SUMMARY

The returning water enters the tank via a horizontal, wide–mouthed "horn," or cone, that slows it almost to a stop, without turbulence, with the consequence that, by virtue of its natural buoyancy (which depends solely on its temperature) it rises or falls to join the temperature-matching stratum.

INTRODUCTION

In most water–type thermal storage systems there is some thermal stratification within the tank, the water being hottest at the top and coldest at the bottom. It is desirable, when introducing hot water to the tank (say, 110°F water from the solar collector), to deliver this water to that particular stratum in the tank that has this same temperature. Mixing of hot and cold water is avoided. Increase in entropy is avoided. Thus, more hot water is available for heating the rooms during the coming night and more cold water is available (at the bottom of the tank) for sending to the collector—which then operates with greater efficiency.

PROPOSED SCHEME

The water is returned via a horizontal, wide-mounted "horn," or cone, situated in the upper part of the tank. Proceeding along the ever-widening cross section of the horn, the water is gradually slowed almost to a stop, without turbulence, and, depending on its

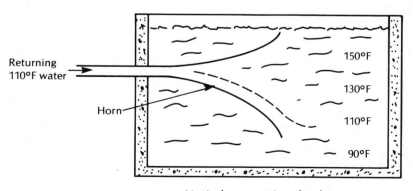

Vertical cross section of tank
and horn for the returning water

temperature relative to that of the water closest to it (in other words, depending on its relative buoyancy) it gradually ascends or descends until it finds itself within that stratum that has the same temperature.

If there is danger that some turbulence will occur, or that the globs of water, in rising or falling to the appropriate level, will overshoot, one can introduce damping (friction) to counter such dangers. Simply install a loose mass of wire, strings, or netting; for example, a loose roll of chicken-wire netting.

Note that the system works automatically, with no moving mechanical parts, and at no cost other than that of the horn—which, of course, may be of very crude construction.

MODIFICATION

Scheme S-50a

Here we provide two horns: one for hot water returning from the collector and one for tepid water being returned from the room radiators. The horn for the latter is situated lower down, because, for this water, the matching stratum is likely to be nearer the bottom of the tank.

DISCUSSION

An account of some experimental studies of thermal stratification in a water-filled tank is presented by Z. Lavan and J. Thompson in an article in *Solar Energy, 19*, 519 (1977). Their results confirm that, ideally, the tank should be tall and slender and the incoming cold water should be introduced at the bottom.

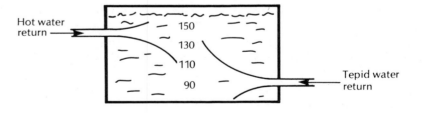

Vertical cross section of tank and horn for the returning water

SCHEME FOR AUTOMATICALLY CHANGING THE CHOICE OF STRATUM [IN A THERMALLY STRATIFIED WATER TANK FROM WHICH WATER FOR ROOM-HEATING IS TAKEN] SO AS TO INCREASE, RATHER THAN DECREASE, THE EXTENT OF STRATIFICATION

Scheme S-51
9/7/78

SUMMARY

Ordinarily, in a house that has a water-type solar heating system, the water that is taken from the storage tank and used to heat the rooms is taken from a fixed stratum in the tank—the top stratum, because it is the hottest. Here we propose that the water be taken from different strata according to the magnitude of the load. For example, the water may be taken from whichever of two strata (at top of tank and near bottom of tank) comes closest to barely matching the load.

Using this load-matching, entropy-minimizing approach, the occupant of the house finds that, often in fall and spring and sometimes in midwinter, the extraction of heat from the tank water actually increases the extent of thermal stratification in the tank. Consequently (a) the ability of the storage system to keep the rooms warm during a subsequent cold spell is increased and (b) the water sent to the collector tends to be colder and, accordingly, collection efficiency is increased.

The repeated changes in stratum from which the water is taken are made automatically. See the first two accompanying figures.

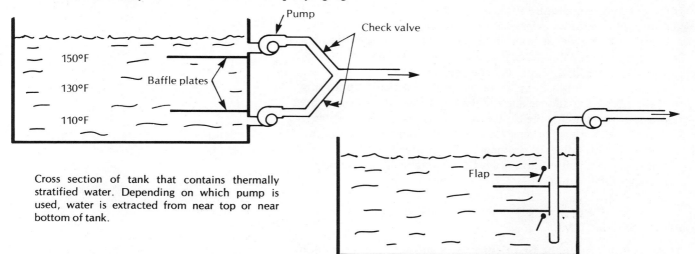

Cross section of tank that contains thermally stratified water. Depending on which pump is used, water is extracted from near top or near bottom of tank.

Here only one pump is used and the stratum from which water is extracted is governed by two flaps.

The third and fourth figures show other automatic systems. The fifth shows an especially simple and cheap system that is controlled manually.

INTRODUCTION

If, in the water–type thermal storage tank of a solar heated house, there is much thermal stratification, it is a shame (wasteful!) to reduce the extent of stratification unnecessarily. If the house occupant can maintain much stratification, he is doubly rewarded: (a) there is more hot water in the top of the tank with which to keep the rooms warm on subsequent cold days and (b) the water that is circulated to the collector (water from the bottom of the tank) is colder and, accordingly, the collector operates at higher efficiency. A physicist would summarize the situation thus: "If, without changing the total amount of energy, you decrease the stratification, you increase entropy, and whenever you unnecessarily increase entropy, you are wasting money."

Thus considerations of economy dictate that, if there is stratification in the storage tank, the rooms need heat, and water is to be taken from the tank and circulated through room radiators, the water taken from the tank should preferably be the *coldest water that will do the job*. For example, if the rooms can be kept warm by circulation of 110°F water, it would be wasteful to use 150°F water.

Because the coldest water in the tank is always at the bottom and the hottest water is at the top, the goal in taking water from the tank is to take the water from the appropriate level or stratum.

As circumstances change, different strata may be appropriate. If the outdoor temperature rises, or the amount of energy received via the south windows increases, or the occupants lower the thermostat, use of a lower stratum in the tank is appropriate. But on a cold night, with the thermostat set high, use of the highest stratum (uppermost water in the tank) may be appropriate.

There are plenty of expensive ways of repeatedly appraising the heat–need of the house and the temperature distribution in the tank and then changing the choice of stratum from which water is withdrawn. Using a set of sensors, a computer, and a servo system, an engineer could solve the problem straightforwardly.

But are there any simple ways of automatically changing the stratum selected? I think so.

PROPOSED SCHEME

Here, as indicated in the first figure, use is made of two extraction pipes set at very different heights within the tank and served by horizontal baffle plates that discourage local vertical motion of the water. A separate water pump is provided for each pipe and separate check valves also. The two pipes are joined together in order to serve the same heat distribution system, which may consist of room radiators, fan–and–coil units, or the like.

The room thermostat used is of a type that has an extra contact. Whenever room temperature falls below 70°F, the first contact is closed and turns on the pump drawing water from near the bottom of the tank. When and if room temperature falls to 65°F, the second contact closes and has the effect of turning off the above-mentioned pump and turning on the pump that draws water from near the top of the tank. Thus, the radiators receive relatively cool water as long as the rooms stay warm enough, and, if they begin to become too cold, hotter water is used instead.

MODIFICATIONS

Scheme S–51a

Instead of using two extraction pipes and two pumps, use a single extraction pipe that has two openings and use a single pump. See accompanying figure. The two openings are at two different heights within the tank and each opening is equipped with a flap that serves as a valve. The thermostat controls the flaps, opening just the lower one when the amount of heat needed is small (room temperature above 65°F) and opening just the upper one when the amount needed is large (room temperature below 65°F).

Scheme S–51b

Use a single pipe and single pump. Drive the pump by means of a two-speed motor which can produce two very different flow rates. When the amount of heat needed is small, the pump is driven at low speed and the flow rate is small. When much heat is needed, the pump is driven at high speed and the flow rate is large.

Near the intake end of the pipe, which is low down in the tank, there is a short, vertical insulated extension pipe, mounted on a small pivot just above its center of gravity. The lower end of the extension pipe is close to (and facing) the open end of the main pipe. The upper end of the extension pipe is in the uppermost region of the water (hottest portion).

When the rate at which water is taken into the main pipe is low, the extension pipe hangs free and plays no role. But when the

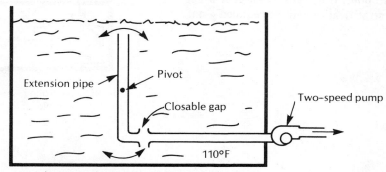

Scheme employing a single two-speed pump
and a pivoting extension pipe

rate is high, the suction near the open end of the main pipe is so great that it causes the lower end of the extenion pipe to swing toward it; the two openings join, with the result that the only way water can enter the system is via the upper end of the extension pipe. In summary, the water extracted from the tank is extracted from low down or high up depending on the flowrate, i.e., depending on whether or not the lower end of the extension pipe becomes coupled to the open end of the main pipe.

Note that at times when the greatest amount of heat is needed, the flowrate is highest and the water is drawn from the uppermost part of the tank.

Scheme S-51c

Here, as indicated in the accompanying figure, there is a single pipe and single pump, and the segment of pipe within the tank has a flexible joint such that the open end of the segment can be swung downward close to the bottom of the tank or upward close to the top. Whether the end of the pipe is low down or high up is controlled by the many-contact thermostat. Various control systems, employing relays, solenoids, or other devices, can be used. One simple arrangement is to have a multi-speed electric motor drive the pump at various flowrates, the flowrate itself controlling the vertical position of the pipe by means of deflector plates, suction orifices, or a counterweight (leaky bucket receiving return water) used with a rope-and-pulley system. The higher the flowrate, the higher the pipe is forced upward.

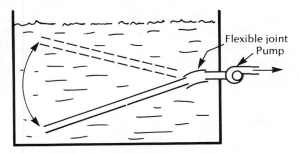

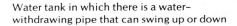

Water tank in which there is a water-withdrawing pipe that can swing up or down

Scheme S-51d

The cheapest system, shown here, is one that is controlled manually. The house occupant raises or lowers the pipe by means of a cable that extends from the pipe to, say, the kitchen. He lowers it when the required rate of heat-delivery to the rooms is low and raises it when a high rate is needed.

The system is simple, durable, and cheap. If the occupant forgets to raise or lower the pipe, no significant harm is done: if he fails to raise it when he should, the rooms will gradually become too cold, which will serve as a reminder; if he fails to lower it when he should, the extent of thermal stratification will be reduced—not a significantly harmful change.

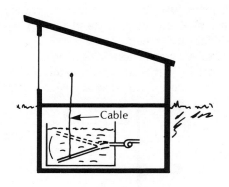

Manually controlled system that is simple, durable, and cheap

AUTOMATIC, OPTIMIZING CHANGE OF STRATUM [IN A THERMALLY STRATIFIED BIN-OF-STONES] FROM WHICH AIR FOR ROOM-HEATING IS TAKEN

Scheme S-52
9/18/78

PROPOSED SCHEME

The following figure shows how, with the aid of special within-bin passages, two special dampers, and a special multi-contact thermostat, the designer can arrange for room air to be circulated just through the lower (cooler) portion of the bin at times when little heat is needed (room temperature above 65°F) and through the upper (hotter) portion of the bin when much heat is needed (room temperature 65°F or below).

When the rooms need only a small amount of heat, Damper #1 is closed and Damper #2 is open. When more heat is needed, Damper #1 is opened and Damper #2 is closed.

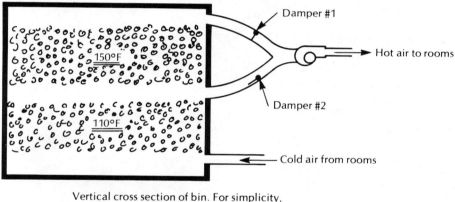

Vertical cross section of bin. For simplicity, the ducts connecting the bin-of-stones to the collector are not shown

MODIFICATIONS

Refine the scheme: use three or four regions of stones and a corresponding number of ducts and dampers.

If the automatic controls would be too expensive, use manual controls. Each damper could be controlled manually by a cable running to, say, the kitchen.

IMPROVING THE PERFORMANCE OF THE THOMASON STORAGE SYSTEM

SUMMARY

Although the Thomason storage system, employing a water tank surrounded by stones, is simple, effective, and low in cost, it has three limitations: (1) airflow past the surface of the water tank is slightly impeded by the stones situated close around the tank, (2) in winter the stones farthest from the tank receive little heat and do little good, (3) in summer when the stones are cooled at night (by a conventional air conditioner) and are used during the day to help cool the rooms, it is not feasible to keep the water tank hot, as for preheating domestic hot water.

Could one improve the performance by leaving a little free space close around the tank and employing a very small blower to continually circulate tank-warmed air through the quantity of stones? I think so.

And would it be possible to isolate the tank from the stones in summer so that two otherwise incompatible purposes could be served simultaneously? I think so.

Whether the cost of such improvements would be justified, I do not know.

I understand from H. E. Thomason that he himself has recently made some improvements in the design. (Many features of his designs are patented.)

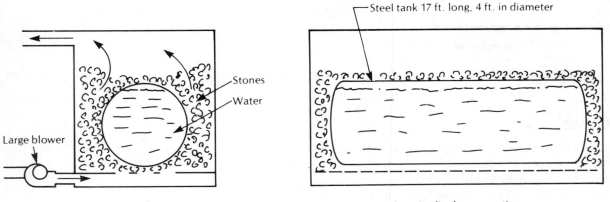

Transverse cross section

Longitudinal cross section

Steel tank 17 ft. long, 4 ft. in diameter

Large blower

Stones

Water

Simple bin containing water-filled tank largely buried in a quantity of stones. For simplicity, many details, for example the water pipes, are omitted

INTRODUCTION

As everyone knows, the typical storage system used by H. E. Thomason includes a long slender horizontal water-filled steel tank that is largely buried within a quantity of 1-to-3-in.-dia. stones. See accompanying figure. Hot water from the collector flows into the tank, heating it. Heat from the tank warms the neighboring stones. When the rooms need heat, room air is circulated through the bin by a large blower. This air picks up much heat from stones close to the tank and a little heat from the tank itself. A detailed description of the system is contained in my book *Solar Heated Buildings of North America: 120 Outstanding Examples*.

A shortcoming of the system is that stones that are far to the side of the tank receive little heat from it. The stones nearest the tank tend to partially insulate it from the stones far from it. In upshot a considerable fraction of the stones may never become warmer than room temperature; accordingly, they contribute little to storage of heat; they are of little use in winter. (But in summer they help considerably in cooling the house.)

The notion that stones that are in direct contact with the tank can "draw heat from it" is largely invalid. Stone is, to a considerable extent, an insulator, not well suited to conducting heat from a hot object. Also, most of the stones that touch the tank make contact with it in very small areas only—practically point contacts. However, infrared radiation helps the flow of heat.

It there some way of speeding the transfer of heat from tank to stones? Is there some way of distributing the heat throughout all of the stones, not just the ones within a few inches of the tank? I think there is.

PROPOSED SCHEME

Install the quantity of stones above the tank, not below it or beside it. Leave space all around the tank so that air can flow freely past it and extract heat from it. Employ a small, low-power blower to draw air from the top of the bin, drive this air downward in a duct, and cause the air to impinge on the bottom of the tank. See accompanying figure. Keep this blower running steadily, 24 hours a day, throughout the four coldest months (at a cost of about $10 in all).

In this way heat is extracted from the tank steadily and is transferred to all of the stones. Thus, more heat is stored in the stones. As a corollary, the tank tends to be cooler and therefore collection efficiency is increased.

If on some occasions the small blower is not running, gravity convection will slowly carry warm air from around the tank to all of the stones. Also, infrared radiation will at all times carry some energy from the tank to the lowest stones, whence the heat will travel slowly upward to the other stones.

The large blower used from time to time to circulate room air through the bin is unaffected by the design changes proposed here. However, it may be necessary to install a check valve in the small vertical duct associated with the small blower in order to prevent reverse (bypassing) flow in this duct when the large blower is running.

(Supporting the stones so high up would be somewhat difficult. Also access to the tank would be poor. In the following scheme these limitations are avoided.)

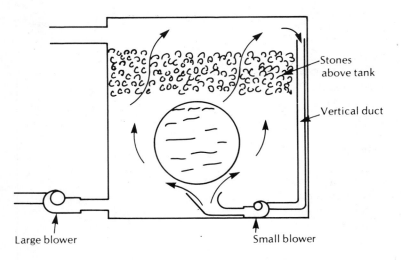

Transverse cross section of proposed system

Scheme S–109a

Here, with the goals of simplifying the support of the stones and improving the access to the tank, we place the stones below and the tank above. Again a small vertical duct and small blower are provided to continually extract heat from the tank and distribute it throughout all of the stones.

An important additional improvement is the provision of an insulating housing for the tank proper. This housing, with its two dampers (one hinged, one sliding; both turned manually in the spring and in the fall), makes it possible to thermally isolate the water tank from the stones in summer, so that (1) the stones can be cooled at night, by means of a conventional air conditioner, so that they can cool the rooms on the following hot day, yet (2) at the same time, the water tank can be kept hot (by continued operation of the roof-top collector) for the purpose of heating the domestic hot water. See figure showing insulated housing.

The performance of such system should be excellent. But would it be enough better to justify the added cost?

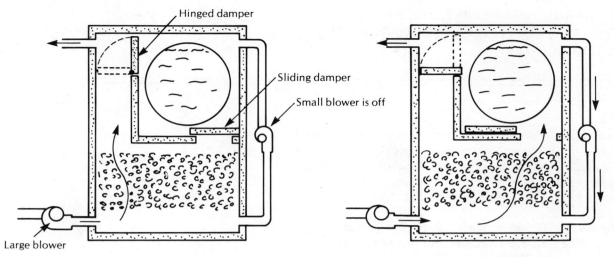

Summer. Air passes through stones only. Stones are cold. Tank is hot and is insulated.

Winter. Large blower drives air upward through stones and then past the water tank. Small blower runs continually to carry heat from neighborhood of tank to the stones

System with (1) tank above stones, (2) small blower to carry heat from tank to stones, and (3) insulating housing that permits keeping stones cool in summer while tank remains hot.

HOW TO MARRY A WATER-TANK STORAGE SYSTEM TO AN AIR-TYPE COLLECTOR: EMPLOY HYPER-INTERFACING

SUMMARY

The accompanying figures show how to marry a water-filled steel tank to the stream of hot air from a roof-top collector. Hyper-interfacing is used. The air is constrained by a set of cowling strips to pass extremely close to the surface of the tank; accordingly, the heat transfer is much (10 times?) greater than when no cowling strips are used.

Does the use of such thin passages lead to very high pneumatic resistance, requiring great blower power? No, because the length of path in the passages is so short and the width of path is great.

Two advantages result from using such water tank instead of a bin-of-stones: (a) 3-fold saving in space, or instead a 3-fold increase in carrythrough, and (b) compatibility with water-type heat distribution systems, for example, a system employing hot-water radiators.

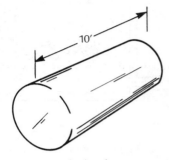

Naked tank

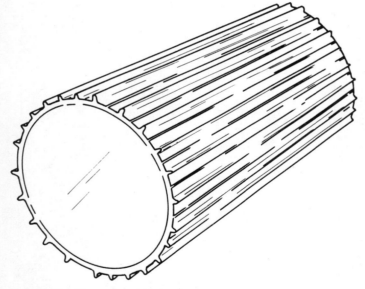

Tank with cowling strips in place

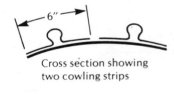

Cross section showing two cowling strips

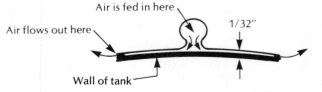

Air is fed in here

Air flows out here

1/32"

Wall of tank

Cross section of one cowling strip, showing directions of airflow

INTRODUCTION

Solar–heated–house designers who specify air-type collectors and air-type heat distribution systems have felt compelled to specify, as storage system, a bin-of-stones, because of the huge heat-transfer area afforded.

Is there anything seriously wrong with a bin-of-stones? Yes: the lamentably small thermal capacity of stones. Consider, for example, granite: at about 122°F granite has a density of 2.65 g/cm^3 and a thermal capacity of 0.173 Btu/(lb, °F), according to a detailed report "Geological Society of America Special Papers, No. 36," 1/31/41, F. Birch, ed., p. 235. If a bin of granite stones is 70% by volume stones and 30% air, the thermal capacity per ft.^{3} is only 32% that of water. The figure 32% is arrived at by the following calculation:

$$\frac{(62.4)(2.65)(0.70)(0.173)}{(62.4)(1)(1)(1)} = \frac{20.1}{62.4} = 32\%$$

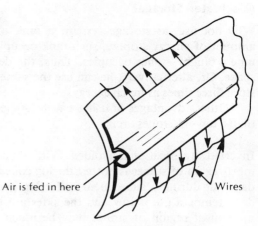

Detail of one cowling
strip and the wires
that maintain a 1/32-in. air-
gap between strip and tank

Air is fed in here / Wires

Should the designer choose a large bin or a small one? He is caught between Scylla and Charibdis:

Scylla: He opts for a huge bin. He uses, say, 100 tons of stones.

The main drawback is that the bin occupies a very large volume, such as 30 ft. × 12 ft. × 7 ft. It preempts ⅓ or ½ of the basement. Another drawback is the cost of building a big strong bin, with plenums at top and bottom, 6 in. of insulation all around, and with no air-leaks. Another drawback is the cost of procuring the stones, shipping them, and washing them.

Charibdis: He opts for a small bin—say about 7 ft. × 7 ft. × 6 ft. (Many solar houses have bins of about this size.)

The drawback is that the carrythrough is very small. If the actual net region of stones is 6 ft. × 6 ft. × 5 ft., the volume is 180 ft.^{3}, which is equivalent to about 60 ft.^{3} (3700 lb.) of water. Thus the amount of heat released when the bin-of-stones cools down from 130°F to 90°F is only (3700)(40) = about 150,000 Btu. This may provide, under various common circumstances in January, enough heat to keep a house warm for only 6 or 8 hours. If Jan. 1 is a sunny day, the bin may become practically "empty" of heat by midnight of that same day.

Use Water Storage?

Why not use, as storage system, a tank of water? For a given amount of energy storage, such tank occupies only ⅓ of the volume a bin-of-stones occupies. Thus, the designer can save much space. Or, alternatively, he can use the same amount of space and store three times as much energy.

Other advantages of using a water-type storage system rather than a bin-of-stones are:

Inverted stratification is avoided. With a bin-of-stones, the uppermost region of stones is hottest during collection on a warm sunny day, but during a subsequent collection period in which the outdoor temperature is low and the skies are somewhat cloudy, the uppermost region of stones may be much colder than, say, the central region. Such inverted stratification interferes with distribution of heat to the rooms and decreases collection efficiency.

Compatibility with a water-type distribution system is achieved. Hot water from the storage tank can be circulated directly to room radiators. The compulsion to use an air-type distribution system is avoided.

Preheating the domestic hot water becomes very simple.

The obvious objection to using a tank of water (to serve an air-type collector and/or an air-type heat distribution system) is that there has been no good way of achieving good heat transfer. If one merely blows the air past the tank, little heat is transferred. One could, of course, install a conventional heat exchanger. But the added complexity and cost are appreciable, I expect.

How nice it would be if the big water-filled steel tank could itself be modified so as to provide a high rate of heat transfer as well as storing much heat. Welding external fins to the tank would be helpful; but (a) welding is expensive and (b) the heat transfer might be increased insufficiently—by only a factor of 2, say.

I expect that a cheaper and more effective way to modify the tank is to employ a set of well-designed, close-fitting cowling strips. Some readers may be familiar with the enormous effectiveness of the special cowlings that were invented in the 1920's by the National Advisory Committee on Aeronautics to improve the air-cooling of radial engines for aircraft. Something analogous might be done for steel tanks.

PROPOSED SCHEME

The figure here shows a scheme which provides greatly improved heat exchange between (a) a horizontal, cylindrical, 4-ft.-dia., 10-ft.-long, water-filled steel tank and (b) a forced stream of air (hot air from collector or cold air from the rooms).

The tank is embraced by a set of 22 cowling strips, each 6 in. wide and 10 ft. long. Each is parallel to the axis of the tank.

Each strip has a central tapered channel and two flanking vanes. Air from a blower enters the large end of the channel and the other end of the channel is blocked. The air escapes laterally from the channel via 1/32-in. spaces between vanes and tank. The spaces are maintained by a 1/32-in.-dia. wire that has been wound helically around the tank. The surface of the tank has been roughened to encourage micro-turbulence of airflow. One manifold distributes air to all of the channels, via their open ends. The task of collecting the air that issues from under the 22 pairs of vanes is performed by the insulating housing that encloses the tank.

The cowling strips, mass produced in a factory, are of cheap thin metal or plastic. They can be held in place on the tank by means of a few straps or belts. Obviously, no exact positioning of the cowling strips is needed.

Installing an input manifold on one end of the tank is easy if one first installs an encircling band of grout or other filler material, to produce a uniform cylindrical tank-end assembly, or band. A simple cup-shaped manifold can now be slipped over this band and sealed to it as suggested by the accompanying sketches. The big insulated housing itself would serve as the output manifold, to collect all the air that has escaped beneath the cowling strips; this air is then circulated, by a blower, to the rooms or to the collector.

Warning: The dimensions and proportions proposed here are based on guesses. The optimum dimensions are probably somewhat different. Presumably they should be determined by means of experimentation.

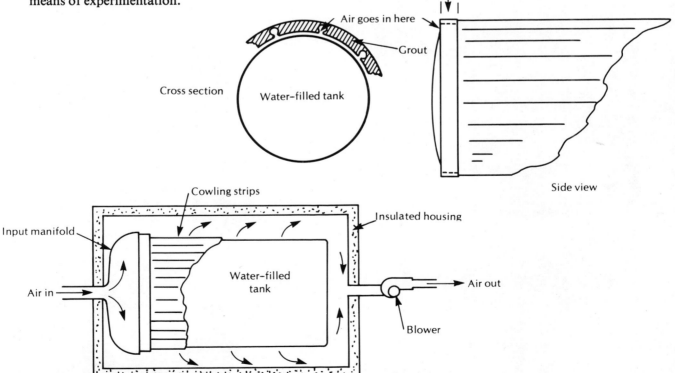

Scheme S-150a

Here there are two such tanks, each equipped with cowling strips. The two tanks are within a single long housing. The blower is incorporated in a partition at the center of the housing and serves the cowling at the right end of the left tank and the cowling at the left end of the right tank.

The great virtue of this system is that the air pressure is approximately the same at both ends of the housing, and may be atmospheric pressure, or near-atmospheric pressure, at both those locations. Thus, there is no great tendency for any air-leak to occur in the equipment (ducts, collector, etc.) that is external to the housing.

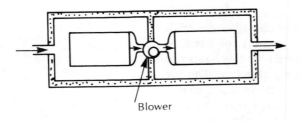

Blower

Scheme employing two tanks. The blower is situated symmetrically between them.

PART 6

Storage Systems Employing Phase Change Materials

INTRODUCTION

As nearly everyone knows, phase change materials (PCMs) can store an enormous amount of heat in a small space. This has great appeal to designers of solar heated houses. However, many such materials have exasperating behavior which can be corrected only at some inconvenience and expense.

A basic difficulty is that so few materials that are cheap and stable undergo a phase change at the desired temperature—say, 90°F, 120°F, 150°F, or other temperature in this general range.

Certain combinations of materials (certain solutions) meet most of the requirements. Some eutectic solutions appear especially promising. Among these are various salt hydrates that were selected and tried out by M. Telkes in the 1940's and in subsequent decades. The best known of these salt hydrates is Glauber's salt, also called sodium sulfate decahydrate ($Na_2SO_4 \cdot 10H_2O$).

EUTECTIC SOLUTION: DEFINITION AND KEY PROPERTIES

I have had great trouble in getting a firm understanding of what a eutectic solution is and what its distinctive properties are. With the help of Professor Paul D. Bartlett, I have arrived at the following definition.

Suppose that a given vessel contains a homogeneous, hot, liquid that consists of two or more compounds (salts, for example) that are mutually soluble in one another. Suppose the (liquid) solution is gradually cooled until crystallization is imminent. If at that time the proportions of the compounds in the solution are such that each compound is saturated with each of the other compounds, the solution is said to be *eutectic*.

Suppose that energy is then continuously extracted from the material. One finds that:

• When crystallization starts, it starts for all of the compounds simultaneously.

- As crystallization progresses, the proportions of the compounds in the liquid phase (likewise the solid phase) remain unchanged.
- When crystallization is complete for one compound, it is complete for all.
- Throughout the period in which crystallization proceeds, the temperature of the material remains constant.

No other combination of these compounds has such a low temperature of crystallization.

ORGANIZATION OF MATERIAL

In the following pages, some drawbacks to the use of salt hydrates are listed, then two commercially promising schemes for packaging such salts—in plastic trays and plastic tubes—are described. Finally, several possible methods of using the salts in large containers are explained.

An attractive method of using salt hydrates in a passive solar heating system is described in Part 1.

FOUR DRAWBACKS TO THE USE OF EUTECTIC SALT FORMULATIONS FOR ENERGY STORAGE IN SOLAR HEATING SYSTEMS

SUMMARY

Thermal storage systems employing eutectic salts, such as the salt hydrates known as Glauber's salt and hypo, can store so much heat in a small space that solar heating system designers everywhere eagerly await the completion of the present development programs and the start of large-scale commercial production.

It should be remembered, however, that even after problems of supercooling, settling out, packaging, leakage, etc., have been overcome, storage systems employing eutectic salts will still have four important drawbacks: (1) single type of use, (2) extreme immobility of the material, which can neither fetch nor deliver heat, (3) tendency of the salt to hamper heat output and heat input, and (4) difficulty of ascertaining, at any given time, how much energy is in the system.

INTRODUCTION

As M. Telkes pointed out 30 years ago, several eutectic formulations can store much thermal energy in a small space. For example, the salt hydrate called Glauber's salt ($Na_2SO_4 \cdot 1OH_2O$) and, likewise, the salt hydrate called hypo ($Na_2S_2O_3 \cdot 5H_2O$) can store about 5 to 10 times as much heat as an equal volume of water can store and 15 to 30 times as much heat as an equal volume of stones can store. Such salt hydrate is ordinarily used in combination with a small amount of nucleating agent and also a small amount of a thickening agent. Glauber's salt melts (or freezes) at about 90°F and hypo melts (or freezes) at about 120°F. Commercial production of shallow boxes, or trays, filled with salt formulation is said to be getting underway—most welcome news.

My understanding is that Glauber's salt costs only a few cents a pound, FOB, if very large quantities are purchased and high purity is not required; in other words its cost is almost negligible compared to the cost of packaging and shipping. Hypo costs only a little more than Glauber's salt.

But are there some inherent drawbacks to the use of such material? There are. I try to list them here, not for the purpose of denigrating the materials but to help provide a basis for a better understanding of how and where to use them.

THE DRAWBACKS

1. A set of sealed, shallow boxes, or trays, filled with a formulation of a salt hydrate suitable for supplying heat to a house in winter cannot be used to help cool the house directly in summer.

For cooling, a different formulation should be used. In other words, any one formulation is likely to have only one type of use. (However, because such a storage system is so small, use of two different systems—one for winter and one for summer—may sometimes be practical.)

Note that water is more versatile. A tank of water can be kept at about 130°F in winter to help heat a house and can be kept at about 50°F in summer to help cool it. The same applies to a bin-of-stones. Thus, such storage systems are able to serve two kinds of purposes, as has already been demonstrated in various solar heated buildings.

2. The salt hydrate will not budge. It stays at all times in its sealed trays. It cannot be fed to room radiators because it might freeze there and block further flow. Thus, the salt hydrate cannot be used to transport heat: it can neither fetch heat from the collector nor carry heat to the rooms. (However, conventional heat-transfer systems employing forced flow of air or water can be used effectively to do the necessary fetching and delivering.)

Note that the water in a tank can be pumped through the pipes of a roof-top collector or through the radiator pipes in a room. That is, water, besides storing heat, can fetch and deliver it. Stones, however, cannot fetch and deliver; but they have, in aggregate, very large surface area, which greatly facilitates heat exchange. The area is about 50 to 500 times that of a thermally equivalent amount of salt-hydrate-filled tanks or trays.

3. When a stream of cool air starts to take heat from a liquid-salt-hydrate-filled tray, a crust of solid material may soon form on the inside faces of the tray and somewhat impede the further extraction of heat. Likewise, if you start to put heat into a tray full of solid-salt hydrate, liquid films form and somewhat impede the flow of additional heat to the portion of the tray-content that remains solid. (However, if—as is usually the case—the containers consist of trays that are thin and large in area, heat transfer proceeds at a rate that is high enough. Incidentally, the thermal conductivity of solid salt hydrate is, I understand, surprisingly high, and is higher than that of the liquid material.)

Note that the water in a tank is highly cooperative. Within the tank, gravity-convective flow occurs automatically and greatly facilitates heat input and heat output. Stones, of course, have no such capacity.

4. It is almost impossible, ordinarily, to ascertain quickly and accurately the amount of energy in a set of salt-hydrate-filled trays. Whether a given tray contains material that is 10% liquid or material that is 90% liquid, the temperature of the tray is practically identical. How, then, is one to find what fraction of the material is liquid and how much energy is stored? To make matters worse, in some trays the material may be largely liquid while in others it may be largely solid. Will not the house occupant be somewhat uneasy if he does not know how much energy is in storage? (Of course, he need not be *very* uneasy, because the furnace stands ready to come on at any time.)

Note that when heat is stored in a tank of water, one or two simple measurements of temperature suffice. Likewise, it is easy to measure the amount of energy in a bin-of-stones.

In principle, one could ascertain what fraction of the material in a set of trays is liquid by (1) use of X-ray diffraction, the refraction being very different for the liquid and solid phases; (2) suspending the set of trays by means of flexible wires and causing the set to oscillate (linearly or rotationally); the rate of oscillation depends slightly—and the damping depends greatly—on the fraction of the material that is liquid. Various sonic methods could be used also. But such methods would probably be expensive. I know of no simple method applicable to a set of many trays.

TWO ADVANCED DEVELOPMENTS IN THE PACKAGING OF SALT HYDRATES

SUMMARY

At least two groups are making rapid progress toward routine mass production of well packaged salt hydrates for use in thermal storage systems: Valmont Energy Systems, Inc., which employs plastic trays, and the Institute of Energy Conversion at the University of Delaware, which employs plastic tubes.

INTRODUCTION

Salt hydrates offer such a striking and highly valuable capability in thermal energy storage, being able to store so much heat in a small space, that many university and industrial groups have been trying to develop practical storage systems employing these materials.

Many difficulties have been encountered and all have been overcome or circumvented by Dr. Maria Telkes in three decades of hard work. In 1978 she was serving as Director of Solar Thermal Storage Development at American Technology University, P.O. Box 1416, Killeen, TX 76541, where she was continuing to push for mass production of durable and cost–effective storage systems employing salt hydrates.

Two groups now making progress toward mass production of such systems are discussed below. Other groups also may be making progress, for example, Solar Inc. of Mead, Nebraska; but I have been unable to obtain much information on their projects.

VALMONT ENERGY SYSTEMS, INC.

This company, situated in Valley, NE 68064, has issued brochures indicating that it is getting ready to produce and sell storage systems that employ Glauber's salt formulation in sealed moisture-proof trays 24 in. × 12 in. × 2 in. Each tray, made of high–density polyethylene, contains about 20 lb. of the formulation. In the upper and lower faces of each tray, there are sets of generous–size, parallel channels to permit free flow of air when the trays are stacked up 15 high. Each pound of salt can store about 100 Btu of latent heat and each tray can store 2000 Btu. If 300 trays are used (suitably stacked) and if an insulating housing is provided, one has a complete system that can store about 600,000 Btu at about 90°F. Information on price and delivery is eagerly awaited. My understanding is that the price may be as low as $12.50 per tray, FOB. The cost of 300 trays would then be almost $4000.

Source of information: "Thermal Storage in Salt–Hydrate Eutectics", M. Telkes and R. P. Mozzer, in *Proceedings of the Am.–ISES August 1978 Conference* in Denver; also brochures from Valmont Energy Systems, Inc.; also personal communications from M. Telkes and from P. B. Popinchalk of Valmont Energy Systems, Inc.

Channeled tray, 24 in. x 12 in. x 2 in.,
containing 20 lb. of salt hydrate formulation

INSTITUTE OF ENERGY CONVERSION AT THE UNIVERSITY OF DELAWARE.

Late in 1978 engineers at this institute in Newark, Delaware, and at cooperating industrial laboratories, were completing the development of salt hydrate storage systems operating at about 90°F or various other temperatures in the range from 39°F to 90°F. An especially low–cost type of packaging is employed: the material is sealed within sausage-shaped plastic tubes 1½ to 2 in. in diameter and 16 in. long. The plastic sheet used is a tough, moisture–proof four–layer laminate. Each tube contains 3 to 5 lb. of salt hydrate formulation. The estimated cost of packaging is said to be about one cent per pound of salt hydrate included. The methods were worked out in conjunction with duPont Company. Late in 1978, moderate–scale pilot plant production was underway. Operated at full capacity, the equipment was capable of packaging 2 tons of material per hour. Negotiations concerning mass production and marketing were underway; several companies were actively engaged in the negotiations.

Sealed tube 2 in. in diameter
and 16 in. long, containing
5 lb. of salt hydrate

In a complete storage system, a very large number of such filled tubes would be used. They would lie horizontally and would be supported by horizontal nets or grills or by slender "rungs" of vertical "ladders" of nylon cord or other material.

IMPROVING THE PERFORMANCE OF A PCM STORAGE SYSTEM BY USING TWO PCMS OPERATING AT TWO DIFFERENT TEMPERATURES

Scheme S–125
9/22/78

SUMMARY

The performance of a phase–change–material (PCM) storage system may be improved if the system is divided into two parts and these employ PCMs having two different operating temperatures. If appropriate airflow patterns are used, the temperature of the air returned to the collector is lower, in many periods in midwinter, and accordingly collection efficiency is higher.

PROPOSED SCHEME

Consider a solar heated house that has an air–type collector, an air–type distribution system, and a PCM storage system. Clearly, the collection efficiency can be increased if the air en route to the collector can be made colder. This can be done by creating some helpful thermal segregation.

    To provide some thermal segregation, one could:

A. Employ a two–part, i.e., two–bin, storage system.
    Part 1 is large and employs a PCM operating at, say, 120°F, and
    Part 2 is small and employs a PCM operating at, say, 90°F.

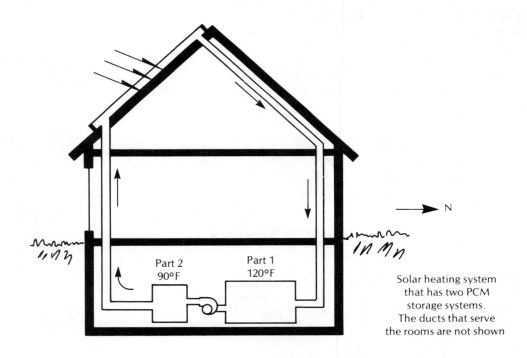

Part 2
90°F

Part 1
120°F

N

Solar heating system
that has two PCM
storage systems.
The ducts that serve
the rooms are not shown

B. Connect the two parts in series, with respect to the collector. Hot air from the collector flows first through Part 1 (the large, hot part), then through Part 2, and then back to the collector.

C. Heat the rooms by means of Part 2 alone whenever this part is capable of keeping the rooms as hot as 70°F. When it lacks this capability, for example on cold January nights or cloudy January days, both parts are used: room air is circulated first through Part 2 and then through Part 1.

Using such system in this way, the house occupant may succeed, during substantial periods in midwinter, in keeping the temperature of Part 2 at about 90°F or lower. Accordingly, the air that is sent to the collector is likewise at about 90°F or lower and therefore collection efficiency is very high.

The various modes of airflow to the rooms are controlled by dampers, which in turn are controlled by a room thermostat that has an extra contact. When the thermostat reads about 70°F, room air is circulated just through Part 2. But when the thermostat reads 68°F or lower, air is circulated through both parts: Part 2 and then Part 1.

DISCUSSION

Once again we are reminded that it is a mistake, usually, to use fixed-value systems. For example, it is a mistake, when employing a bin-of-stones, to use stones of the same general size throughout the bin. It is a mistake, when employing a Trombe wall, to have the wall consist of one single piece, or to have many pieces lying in one single plane, or to have all the pieces of the same thickness. Here we show that it is a mistake, in designing a PCM storage system, to have a single operating temperature. In many ways, a solar heating system is improved by building into it a variety of key dimensions and parameters capable of providing, together, versatile and flexible performance. By way of illustration, if a builder asks you to go to a hardware store and buy six screw-drivers, do not choose six identical ones. A variety will be far more useful.

So, if a designer of a PCM storage system is wondering whether to use Glauber's salt with an operating temperature of 90°F or hypo with an operating temperature of 120°F, he might consider using both.

Of course, using both entails added complexity and cost. Whether the game is worth the candle depends on many circumstances. For large installations, use of two formulations might pay off. For small installations it may not.

PCM-FILLED ROTATING DRUM THAT IS IMMERSED IN A WATER-FILLED TROUGH

SUMMARY

The phase change material (PCM) is contained in a sealed, steel, cylindrical, horizontal drum that lies within a water-filled trough and is continually rotated slowly by a 1/15–HP electric motor in order that settling-out problems can be avoided. The water in the trough serves as interface between collector and drum and between room-radiator-system and drum and also provides a valuable buffer function. Change in phase of the PCM within the drum changes the pressure there, and accordingly a pressure gage that measures pressure within the drum provides a measure of the energy content of the PCM.

Scheme S–130a is simpler. A slim water-filled jacket is used instead of a water-filled trough.

PROPOSED SCHEME

A. *Design* The heart of the proposed system, applicable to a solar heated house that has a water-type collector, is a sealed, steel, cylindrical, horizontal drum 4 ft. in diameter and 9 ft. long, filled about 85 or 90% full with $Na_2S_2O_3 \cdot 5H_2O$ or other PCM such as has been tried out successfully by M. Telkes. Some nucleating agent (borax, for example) is included and some anti-corrosion agent. The phase-change temperature is about 120°F. The drum is totally immersed in 3 tons of water contained in an insulated rectangular trough (about 10 ft.× 5 ft.× 5 ft. in outside dimensions) made of wooden braces, plywood, and a waterproof liner. The drum rests on four near-frictionless non-corroding rollers. One pair of rollers, fixed to a common shaft, is rotated continually by a 1/15–HP gear-reduction electric motor; thus, the drum itself rotates at, say, 4 rev. per hr. A pressure gage affixed to one end of the drum shows the pressure within the drum. Most of the equipment is assembled at a factory, delivered by a truck, and (by means of a hoist on the truck) is lowered into place in the base of the house in question.

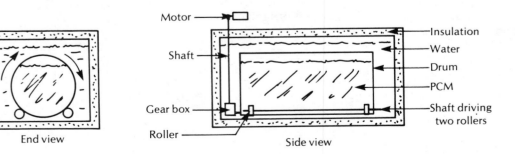

End view

Side view

B. *Operation* Hot water from the collector is circulated to the trough and delivers heat to it and (indirectly) to the PCM. When the rooms need heat, water in the room radiators is circulated to the trough; thus, taking heat from it and (indirectly) from the PCM. The motor keeps running and keeps the drum rotating.

The pressure gage continues to indicate the pressure within the drum. The pressure correlates directly with the volume of the solid-and-liquid within the drum, which in turn correlates with the amount of latent heat in the drum.

Inasmuch as the quantity of PCM used has a volume of about 104 ft.^{3}, a density of about 104 lb./ft.^{3}, a total mass of about 11,000 lb., and a latent-heat of phase-change of about 90 Btu/lb., the total amount of heat liberated when the PCM changes from liquid to solid is about 1,000,000 Btu, i.e., enough to keep a moderate-size, well-insulated house in Massachusetts warm for two days in a typical sunless period in January.

C. *Discussion* Because the drum is immersed in water, the downward force exerted on the supporting rollers is small.

Because the rollers are situated close to the ends of the drum, no appreciable deflection of the drum walls is produced.

Because the roller system is nearly frictionless and the rate of revolution of the drum is low, a low-power electric motor suffices. It uses about 2 kWh (about 10¢ worth) of electricity a day.

Because the drum is rotating, there is no stagnant layer of water adjacent to the outer surface of the drum. Circulation of water from collector or room radiators helps insure that there is no stagnant layer here.

Likewise there is, usually, no stagnant layer of material adjacent to the inner surface of the drum wall. Exception: when drum content is 80 (?) to 100% solid.

This serious limitation of the system must be recognized: the area for heat transfer from PCM to the water in the trough (or vice versa) is small. The amount of heat that can be transferred per 24-hour period may be embarrassingly small. (By way of contrast: in a system that employs hundreds of small trays, the surface area for heat transfer is much greater—of the order of 50 times greater. However, the rate of heat transfer per unit area may, in such system, be significantly lower.)

A sudden supply of heat from collector to trough may be accepted efficiently inasmuch as the water in the trough acts as a buffer: the 3 tons of water can take up a sudden pulse of 100,000 Btu and become only 17 F deg. hotter. Likewise 100,000 Btu of heat can be extracted from the water by the room radiator system in a short time interval, with only a modest drop in temperature of the water. Therefore, the transfer of heat from trough to drum, or vice versa, can be permitted to occur slowly. In several senses, the 3 tons of water serves as an ideal interface between (1) drum contents and (2) collector and/or room radiator system.

Leakage of PCM is almost out of the question, inasmuch as the material is contained in a single, sealed, factory-built container of heavy-gage steel. Corrosion is no problem: a corrosion inhibitor is used.

The storage system is compact: it contains almost no waste space.

By glancing at the presure gage, the owner can at any time estimate the amount of latent heat in the drum.

There may be no need to include, in the drum contents, an anti-settling agent: the steady rotation of the drum discourages permanent or harmful settling and encourages re-intermingling of solid and liquid components.

Is there need for an anti-supercooling agent (nucleating agent)? Perhaps not. The slight turbulence maintained within the drum by the steady rotation may discourage supercooling. Also, an arrangement may be made to stop extracting heat from the drum when the fraction of solid material is reduced to, say, 5%; that is, an ample modicum of crystals (for seed) could be left in the drum at all times. But, in any event, adding a few percent of borax is simple.

The pipes bringing water to or from the trough come in from the top. No water-tight seals (i.e., seals at sides or bottom of the trough) are needed.

My understanding is that the amount of hypo required, about 11,000 lb., would cost about $300 FOB if low-purity material is acceptable and is to be distributed to a large number of users.

Would the proposed system be cost-effective? Would it be more cost-effective than a system employing PCM-filled trays such as are now being readied by two or three different concerns? I would expect the system proposed here to be less cost-effective for small installations, but it might be more cost-effective for large installations. (Note: If a 20-lb. tray were to cost about $12, and were to store 2000 Btu, then a quantity of trays capable of storing 1,000,000 Btu would cost about $6000.)

D. Minor Modifications

1. Make the drum much smaller and the trough likewise. A closet-size system would store enough heat for about one sunless midwinter day. A smaller-diameter drum has a better surface-to-volume ratio.

2. When the manufacturer delivers the storage system, let him deliver it hot—already full of energy.

3. Fill the drum only 55 or 60% full—so that it will float. Then no supporting rollers are needed. The drive could consist of a tiny perimeter-friction-drive motor. But one regrets wasting 40% of the volume of the drum.

4. Provide for off-peak electric heating of the drum whenever its energy content is in danger of becoming very low.

5. Include, within the drum, several free-sliding, sharp-edged, steel disks that will loosen material threatening to adhere to the drum wall.

6. Add 5 or 10% more water to the salt hydrate formulation. Some crude experiments performed by me in 1976 with a eutectic hypo formulation and with a formulation to which extra water had been added indicated that, when there was a little extra water, the material, when cooled, behaved much better: the material was, typically, in the form of a slurry, rather than consisting of a hard and tenacious mass of crystals. A slurry can be stirred but a tenacious mass of crystals—somewhat resembling solid concrete—cannot. The shift in melting point is small and the decrease in thermal capacity is small. Some investigations made by E. R. Biswas of the University of California suggest that the behavior of a Glauber's salt formulation is considerably improved by adding a little extra water. See *Solar Energy, 19*, 99 (1977).

MAJOR MODIFICATION

Scheme S-130a

Here is a simpler version of the above. Use a long slender drum. Provide a jacket that embraces most of the cylindrical surface of drum. Between the jacket and the surface of the drum there is a thin region in which water circulates: water from the collector and/or water from the room radiators. The drum and jacket rotate together, in sealed combination. Each time this combination has executed ¾ of a turn, the direction of rotation is reversed; there is a reversing device applied to the electric motor or its gear drive system. The jacket has inlet and outlet tubes serving the collector or the set or room radiators. The tubes do not get "wound up," because the direction of rotation of the drum keeps reversing; the net wind-up never amounts to as much as one turn. The entire storage system is enclosed in thermal insulation; this may be affixed to the tank and jacket, turning with them; or it may be separate and stationary.

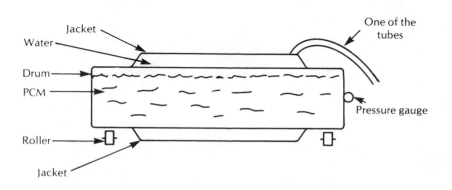

Notice that there is no trough. The jacket serves in place of it. Also, no large amount of water is involved. The rollers, gears, etc., are in air. The system is simple and compact. But it has no large volume of buffering water—unless a separate buffer-tank is provided. And, again, the total surface for heat transfer from or to the contents of the tank is much smaller than one might desire.

PATENT SITUATION

Late in 1978, Broad Corp., of 73 Tremont St., Boston, MA 02108, obtained a U.S. patent (No. 4,117,882) involving some of the main ideas of Scheme S-130 and S-130a. I am listed as inventor. Broad Corp. owns the patent.

I learned early in 1978 that General Electric Co. had developed a salt-hydrate thermal storage system employing some of the same ideas presented above. A rotating drum is used.

PCM-FILLED STEEL TANK THAT FURNISHES HEAT THROUGH THE UPPERMOST SURFACE ONLY, THIS SURFACE BEING PERIODICALLY FREED OF ADHERING CRUST OF PCM BY BRIEF APPLICATION OF VERY HOT WATER

Scheme S-136
8/13/78

SUMMARY

Heat is extracted from the sloping top surface of a PCM-filled steel tank by water that trickles down this surface. Gradually a layer of solid PCM forms on, and adheres to, the underside of the tank-top and slows the extraction of heat. Such layers, or crusts, are periodically released by applying much heat to the tank-top for about 30 seconds. The crusts then fall to the bottom of the tank, leaving the tank top again in contact with *liquid* PCM.

The amount of latent heat in storage at any one moment can be ascertained by means of a specially calibrated pressure gauge that senses the pressure of the air trapped at the top of the tank.

Heat *input* to the tank is via the steel bottom of the tank.

Several modifications are described.

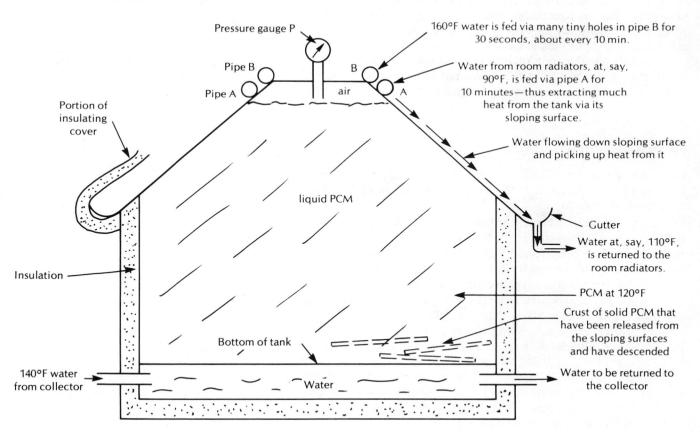

Vertical cross section of proposed storage system.

208

The main heat *input*, from a solar collector, is at the bottom. The main heat extraction, or *output*, is via the sloping metallic surfaces of the tank top. Water from the room radiators is fed via pipe A; this water trickles down, is caught by the gutter, and, at a higher temperature, returns to the room radiators. At about 10-minute intervals, the flow via pipe A is halted and, for about 30 seconds, flow of very hot water via pipe B is initiated—heating the sloping surfaces and the uppermost 0.001 inch of PCM crust in contact therewith, and producing a little melting and thus releasing the crust (which may be, say, 1/8 inch thick).

INTRODUCTION

Various simple schemes for employing a phase change material (PCM), for example hypo (sodium thiosulfate pentahydrate, $Na_2S_2O_3 \cdot 5H_2O$), are defeated by the tendency of the PCM to solidify on—and tenaciously adhere to—the surface through which heat is extracted. If a PCM layer 1 or 2 inches thick builds up on that surface, the rate of extraction of heat becomes extremely small.

In principle, one could remove the adhering layer by mechanical means, e.g., by means of a mechanical scraper (say, a device like a windshield wiper). But the adhering layer is so hard, so tenacious, and so inaccessible that the strategy is likely to fail. Also, the scraper itself may become immobilized when practically all of the PCM in the tank in question is solid.

Another possible solution is to coat the metallic surface with a material to which the solid PCM will not adhere. But to find a material that would perform well in such role for ten years may be difficult.

PROPOSED SCHEME

Here I propose a radically different solution. I thought up the basic features of this scheme in 1976. See my memoranda of 7/22/76, 7/24/76, and 7/26/76.

In the proposed scheme, the heat extraction surface utilized is the uppermost (top) surface. As heat is extracted via the sloping steel plate here, a crust of solid PCM will form on the underside of this plate. Because the solid PCM is more dense than the liquid-phase PCM, the crust has negative buoyancy: it tends to sink. If it can be mechanically released from the metallic surface in question, it will sink toward the bottom of the tank. The PCM that, immediately thereafter, is in contact with the surface in question is *liquid* PCM, and accordingly, the transfer of heat resumes at a high rate.

The solid layer of PCM that adheres to the underside of the metallic tank top can be released by briefly flowing very hot water along the *upper* surface of the tank top. If the PCM is hypo, one might flow 160°F water over the tank top for 30 seconds to loosen (melt) the bond between steel and PCM.

Heat extraction can proceed for, say 10 minutes, and then the resulting adhering crust can likewise be released by a 30-second flow of very hot water.

The very hot water obtains its heat from, say, a concentrating-type solar collector; or from an oil or electric heater. My best guess is that the amount of heat extracted from the tank is 50 times the amount of heat needed for releasing the adhering crusts. In any event this latter heat is not wasted: it ultimately joins up with the normally extracted heat.

Main Heat Flow to Tank

The main heat flow to the tank as a whole (e.g., the flow from the solar collector) is via the base of the tank. Usually there is a quantity of solid PCM resting on the bottom; this takes up the heat furnished and melts. The resulting liquid is less dense than the solid material; thus, the liquid tends to rise and additional solid material tends to descend and rest on the bottom. Eventually, as more and more heat is supplied, all of the PCM becomes liquid. The overall dynamics, or logistics, of the within-tank processes are highly favorable.

At the top of the tank there is a generous-sized airspace which accomodates the volume increase that occurs as the PCM is melted.

When heat is extracted from the tank, and more and more of the PCM solidifies, the volume of the PCM decreases and the volume of the airspace at the top of the tank increases. The volume change is of the order of 5% of the total volume of the tank, i.e., it is fairly large.

System for Indicating the Amount of Energy in the Tank

At the top of the tank there is a pressure gauge which indicates the pressure of the air that is trapped there. This correlates, obviously, with the fraction of PCM liquid, which in turn correlates with the amount of energy stored in the PCM. If the gauge is suitably calibrated, it can indicate *amount of energy in storage* directly—provided that the PCM is at, or approximately at, its fusion temperature, i.e., when the energy at issue is latent heat of fusion, not sensible heat.

DISCUSSION

The surface areas for heat input and output are relatively small, unforunately. Also, the system is comwhat complicated: it employs two kinds of trickling water— *very hot* and *not-so-hot*—flowing alternately. But it is also simple in some ways: inside the big tank ther are no moving parts. Nothing moves except liquid and solid PCM, and such motions are self-initiated and helpful.

During heat input and also during heat output the pathlength of heat-flow is extremely short: the path is only as long as the metal sheet is thick, or perhaps 1/8 in. longer (during heat output) if the adhering solid crust becomes 1/8 in. thick.

It is interesting that if all of the PCM is liquid, and heat extraction is started, and a portion of the PCM begins to solidify, this portion—being uppermost—tends to be the least dense portion, i.e., the portion richest in water, the portion having the *lowest* melting point. I cannot see anything unfortunate about this, however.

The successive descents of plates of solid PCM help stir and homogenize the liquid PCM. This is, of course, helpful.

Many of the comments, design variations, etc., mentioned in my report of 7/17/76 on Scheme S-134 and my report of 7/19/76 on Scheme S-135 apply here also.

MINOR MODIFICATIONS

1. Provide a buffer tank for the output heat from the main tank. The buffer tank, which may have a volume of 250 gallons, receives heat whenever this tank is colder than 115°F and there is no immediate need for sending heat to the room radiators. The main tank provides the main storage, and the buffer tank provides a small amount of quickly available heat. Thus, the process of extraction heat from the main tank can continue 24 hours a day (assuming there is plenty of heat in this tank), even if the radiators need hot water only 8 hours a day.

2. Replace the pressure gauge with a simple float gauge. But provide a 10-watt electric heater in the float proper in order that the float will always find itself in contact with *liquid* PCM, even when 99% of the PCM is solid. The higher the float gauge, the greater the amount of heat in storage.

3. The occasional bursts of heat required to release the solid layers of PCM could be supplied by a jet of steam, by a set of heat-lamps, by high-frequency electromagnetic fields, or by a set of fine nichrome wires in which an electric current flows for 30 seconds.

4. Mechanical vibrators could be used to help release or dislodge the adhering PCM crusts.

5. One could use, as PCM, Glauber's salt instead of hypo. Various other PCM's could be used.

MAJOR MODIFICATIONS

Scheme S-136a

One could dispense with the airspace in the top of the tank: one could have the tank at all times full to the very top with PCM, so that at all times the entire sloping top surface would be available for extraction of heat from the tank. The changing volume of the

PCM could be accomodated by making the tank bottom slightly flexible and applying a steady upward force to it—a force sufficient to keep the tank full to the top with PCM at all times. The central portion of the tank bottom might be 1 inch higher when the PCM is all solid than when it is all liquid. A measure of the height of the middle portion of the tank bottom would be a measure of the amount of heat that had been extracted from the PCM. The upward force mentioned could be applied by coil springs, or by hydraulic or pneumatic pressure.

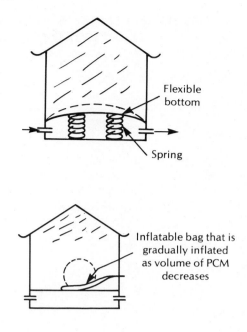

Flexible bottom

Spring

Scheme S-136b

Instead of employing a flexible tank-bottom, employ a fixed tank bottom and install immediately above it an inflatable bladder or bellows. A high enough pressure would be maintained in the bladder so that, at all times, the tank is filled to the top with PCM.

Inflatable bag that is gradually inflated as volume of PCM decreases

Scheme S-136c

Provide a mechanical stirring device of a type that will not become trapped in the PCM as heat is extracted from it and the liquid material changes to solid. The stirrer consists of a ladder-like structure with rigid heavy steel rungs and flexibly linked vertical rods. It is rotated by a vertical driving shaft that passes through the top central part of the tank. As heat is extracted from the tank and the height of the pile of crusts increases, the bottom rungs, while continuing to turn about a vertical axis and drag along the surface of the pile, gradually rise up so as always to remain on top of the pile, free of any strong entanglement.

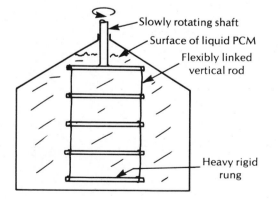

Slowly rotating shaft
Surface of liquid PCM
Flexibly linked vertical rod

Heavy rigid rung

PCM liquid, ladder extended

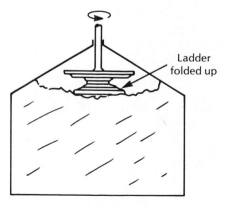

Ladder folded up

PCM solid, ladder folded up. It is much twisted and the rungs drag along the top of pile of crusts.

PCM THERMAL STORAGE SYSTEM EMPLOYING MANY LONG THIN HORIZONTAL STEEL ROTATING PCM-FILLED CYLINDERS FLOATING IN A TANK OF WATER.

Scheme S–137
8/13/78

SUMMARY

The proposed system employs many long thin horizontal steel cylinders, about 60% filled with phase change material (PCM). The cylinders float in a tank of water and are slowly rotated. Heat input (and likewise heat output) is via the tank of water. The heat input (or output) surface is gratifyingly large.

INTRODUCTION

Several of my PCM thermal storage schemes have some drawbacks. The present scheme avoids some of these drawbacks.

PROPOSED SCHEME

The PCM is contained in many long, thin, horizontal, thin-walled cylinders which are, say, 1 ft. in diameter. Each cylinder is about 60% filled with PCM—say hypo ($Na_2S_2O_3 \cdot 5H_2O$); thus each is barely buoyant in water. The cylinders are arranged in two layers, one above the other. Each layer includes several parallel cylinders.

All of the cylinders are within a large, insulated tank of water. Several of the cylinders are connected by belts to pulleys, or sheaves, driven by a small electric motor; these cylinders are turned at the rate of one revolution per 4 minutes. Some cylinders are driven by contact with others.

Heat is delivered to the cylinders via the water in the big tank. Likewise, heat is extracted from the cylinders via the big tank. A pressure gauge along the axis of one cylinder shows the pressure in that cylinder and provides a measure of the amount of latent heat in the cylinder.

Various other design features, considerations, etc., of my other schemes apply here also.

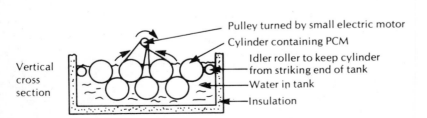

Vertical
cross
section

Pulley turned by small electric motor
Cylinder containing PCM
Idler roller to keep cylinder from striking end of tank
Water in tank
Insulation

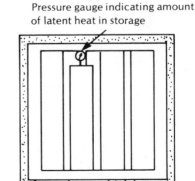

Pressure gauge indicating amount of latent heat in storage

Plan
view

PCM THERMAL STORAGE SYSTEM EMPLOYING A STACK OF PCM-FILLED MATTRESSES BETWEEN WHICH ARE THIN WATER-FILLED MATTRESSES THAT ARE PERIODICALLY INFLATED

Scheme S–137½
8/14/78

PROPOSED SCHEME

Employ several flexible plastic bags, bladders, or mattresses that are filled with hypo or other PCM. The mattresses are sealed and arranged one above another, with spaces between. Each mattress might be several inches in height and many feet in length and breadth.

In the spaces between the PCM mattresses, there are water-filled mattresses each of which has a large inlet tube, an outlet tube, and various hydraulic restrictive devices such that, (a) when E, the excess of inlet pressure over outlet pressure, is small, each water-filled mattress has reasonably uniform thickness, but (b) when E is periodically made very large, the shape of the mattress is greatly distorted, the mattress becoming much thicker in some regions than others.

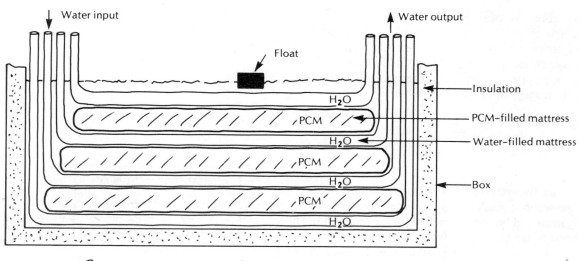

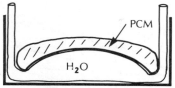

Sketch showing one PCM-filled mattress
greatly distorted by inflating the
underlying mattress with
a large amount of water

The water that flows within the water-filled mattresses may carry heat to the PCM mattresses or may extract heat from them. Thus, we have here a dual-purpose heat-exchanger that has very large heat-transfer surfaces as well as very large latent heat storage.

Periodically, E is varied greatly (by means of timers, water pump, and other devices) so as to greatly alter the shapes of the PCM-filled mattresses. Altering their shapes has the helpful effects of (a) somewhat stirring, or mixing, the contents, to help eliminate any stratification or other tendency toward inhomogeneity, and (b) breaking any crusts of solid PCM forming within those mattresses—crusts that would impede the extraction of heat from those mattresses.

The entire set of mattresses may be enclosed in a large, insulated box that has a waterproof liner. Whenever the flow of water to the water-filled mattresses ceases, and the volume of these mattresses has a fixed minimum value, the overall height of the entire stack of mattresses depends on the fraction of the PCM that is liquid, being greatest when it is 100% liquid. Thus, a device that indicates the overall height serves, in effect, as an indicator of amount of latent heat in storage. If enough water is added to the assembly so that water covers all of the mattresses, a float resting on the water indicates, by its height, the quantity of stored energy.

DISCUSSION

The system looks good. It can be of almost unlimited length, breadth, height. Little or no "high tech" is involved. No metals are used. There is no corrosion problem. The heat exchange surface area is great. Storage capacity is great. The PCM is stirred. Crusts are broken. The water provides excellent thermal buffering. The amount of energy in storage can be determined easily.

MODIFICATIONS

Scheme S-137½a

As above, except omit the water-filled mattresses and immerse the PCM-filled mattresses in a single body of water. Water is fed at one end of the big insulated box and flows out at the other end. The PCM-filled mattresses are kept apart (so that water can circulate between them) by means of flexible cables or hammocks or nettings.

Periodically the PCM-filled mattresses are distorted by tilting, twisting, or bending the box, or by other means. Thus, their contents are stirred and crusts of solid PCM are broken up.

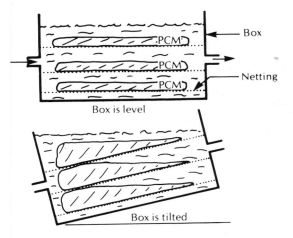

Scheme S-137½b

As above, but balance the box on a central pivot and cause the tank to rock, or tilt, every few minutes (when water is being supplied to the box) by means of a small water tank that rests on one end of the box and is continually and slowly being filled by the stream of water from the collector or the room radiators but which periodically empties suddenly (by siphon action) when the small tank becomes full. Thus, the tilting occurs automatically whenever heat is being added to, or taken from, the PCM-filled mattresses, and no external source of power is needed, other than the normal supply of water.

If tilting is necessary only very infrequently, the occupant of the house could tilt it himself with the aid of a crowbar or ropes and pulleys.

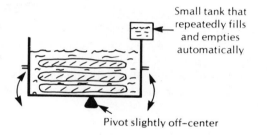

Scheme S–137½c

Here the PCM-filled mattresses are in a single body of water, which in turn is within a box that remains in fixed position. Near the east end of the box there is a vertical array of several small plastic bags, interleaved between the PCM-filled mattresses so that

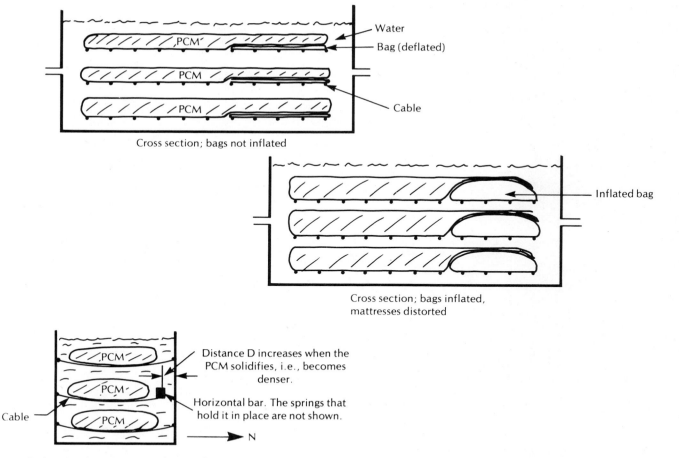

Cross section; bags not inflated

Cross section; bags inflated, mattresses distorted

End-view cross section, looking west

each such mattress is in contact with (partly rest on) one bag. Periodically, all of the bags are inflated, with the result that the quantities of PCM in the east ends of the mattresses are driven westward within the mattresses, stirring the contents of the mattresses and breaking crusts of solid material. The bags may be inflated with air or with water; in the latter case the water may be supplied by a special pump or (with the aid of a timer and appropriate controls) by the same pump that circulates water through the box.

Measuring the amount of stored latent energy in the PCM is accomplished by measuring the combined forces exerted by the north ends of the cables that support the central PCM-filled mattress. When the PCM in that mattress is solid, the mattress volume is about 5% less than when the PCM is liquid; thus the buoyant force is reduced and the tension in the cables is greater. The cable ends in question are secured to a spring-loaded horizontal bar, and the horizontal deflection of this bar is a measure of the percent of PCM (in the mattress in question) that is in liquid state.

PART 7

Domestic Hot Water Systems

INTRODUCTION

This book deals mainly with solar heating systems for buildings, i.e., for space heating. But in this part we deal with domestic hot water.

Many solar domestic-hot-water systems have been described in popular periodicals, including *Mother Earth News, Alternative Sources of Energy, Popular Science Monthly* and in various newsletters and bulletins published by local solar energy societies. No attempt is made here to review those systems.

A few recently invented schemes are described in the following pages.

SOLAR DOMESTIC-HOT-WATER-SYSTEM EMPLOYING A CYLINDRICAL REFLECTOR THAT DIRECTS RADIATION UPWARD TOWARD A BLACK, HORIZONTAL, CYLINDRICAL TANK WITHIN AN INSULATING CANOPY

Scheme S–165 rev.
8/5/77

SUMMARY

The heart of the system is a large, black, horizontal, cylindrical, galvanized-iron tank that serves both as collector proper and storage system. It is situated outdoors adjacent to the south side of the house. Solar radiation reaches it from below via a crude cylindrical reflector. The tank is well insulated at all times—by 6 in. of urethane foam on top and sides and by a 4-in. stabilized air-film on bottom. Inasmuch as freeze-up cannot occur, no antifreeze is used and no heat exchanger. There are no sensors, indicators, pumps, or controls, and no electric wiring. Doubly beneficial thermal stratification occurs within the tank. Tank warm-up starts the instant radiation is received. The device is virtually vandal proof. Installing the device is very simple. The collector tilt is adjusted manually once a month.

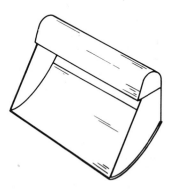

Perspective view

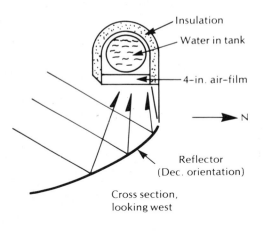

Cross section,
looking west

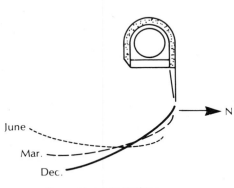

Successive orientations
of the reflector

DESIGN GOALS

The main design goal was to avoid the need for *two* large and costly items: collector and storage system. *One* large and costly item could serve both purposes. Other goals were to eliminate the needs for antifreeze, heat exchanger, sensors, indicators, water pump, controls, and electrical wiring. Such eliminations increase efficiency, simplicity, and reliability and reduce cost.

PROPOSED SCHEME

In the proposed scheme, which has much resemblance to Scheme S-10 of 3/22/73 (see p. 159), a single horizontal, cylindrical, galvanized-iron tank is used as collector proper and as storage tank. The bottom of the tank, which absorbs radiation, has a selective black coating. The tank is 12 ft. long and 18 in. in diameter. It contains 30 ft.^{3} of water; i.e., 150 gal., or 1250 lb. It is enclosed (except at the bottom) in 6 in. of urethane foam (U = 0.03) protected by a weatherproof covering. The bottom is insulated by a 4-in.-thick air-film defined by two sheets of glass 4 in. apart; because the hottest part of the air-film is the top part, and hot air tends to rise, no convection currents occur in this air—the thermal stratification is stable, and the U-value of the 4-in. air-film is comparable to (a guess) that of an equal thickness of fiberglass. Thus, the U-value of the tank enclosure as a whole is about 0.04. Below the tank there is a 14-ft.× 6-ft. cylindrical, concave reflector with an anti-tarnish-coated aluminum face (aluminized mylar? King-Lux aluminum?) that has a reflectance of about 85% when new and somewhat less a few years later. Reflective ends are provided also. The reflector is tilted so as to direct the solar radiation upward toward the tank bottom during at least the 5-hour midday period. The occupants adjust the tilt every month or so. The tank is supported by a sturdy stand. A well-insulated pipe carries city-main water to the lower part of the tank and another such pipe carries hot water from the top of the tank to the input pipe serving the existing hot-water tank in the house.

OPERATION

Operation of the system is automatic and uneventful. There are no sensors of any kind—no sensors of temperature or radiation level or time. There are no controls of any kind—no valves, pump, switches, no electric wiring. Nothing moves, other than the water itself, propelled by city-main pressure, whenever anyone in the house draws hot water. The system is passive—the sun shines, and energy is absorbed.

PERFORMANCE

On a sunny day in December the collector receives about 12 ft.× 4 ft.× 250 Btu/(ft.^{2} hr) = 12,000 Btu/hr, or about 60,000 Btu per sunny day. About 65% of this, or about 40,000 Btu, is absorbed by the tank. This is enough to raise the temperature of the 1250 lb. of water 40,000/1250 = 32 F deg.

If the tank is 50 F deg. hotter than the outdoor air, the 24-hr. conductive heat-loss via the 60-ft.^{2} of enclosing insulation (U = 0.04) is 60 × 50 × 0.04 × 24 = 3000 Btu, i.e., enough to cool the tank by 3000/1250 = 2.4 F deg., if there are no other inputs and outputs. The loss amount to ∼ 7½% of the sunny-day input and about 12% of a typical day's input. The radiative loss (downward) is about equally great.

The performance in March is better, inasmuch as the length of day is greater and the tilt of the reflector, if correct at noon, remains close to correct all day. Also, the ambient temperature is higher. In summer the performance is excellent inasmuch as the ambient temperature is high and the temperature of the city–main water is high.

I estimate that for the year as a whole the system will deliver about 45,000 Btu per sunny day and 25,000 Btu per average day. This is about 9,000,000 Btu/yr. or 2700 kWh, worth (at 6¢/kWh) about $160.

RELATIONSHIP TO EARLIER SCHEMES

The present scheme is somewhat like the space–heating schemes S-10 and S-25 reported by me in March and May 1973 and even more like scheme S-162-h of 6/28/77. The scheme has some resemblance to the Falbel Fes–Delta collector and to many other types of equipment (such as those by Baer and by Khanh) that employ crude reflectors adjacent to cylindrical tanks. Early in 1977 Ian Cannon of Worcester Polytechnic Institute built a system patterned on the present scheme S-165-rev. See also an article by Conrad Heeschen in *Proceedings of the 2nd National Passive Solar Conference,* March 1978, p. 632.

ACKNOWLEDGMENTS

I am indebted to Raymond Bliss for discovering, and firmly pointing out, that the original scheme (S-165) contained an enormous flaw: I had forgotten about radiation loss and the need for using glazing of glass and the desirability of using a *selective* black coating on the tank.

MINOR MODIFICATIONS

Use several short tanks instead of one long one. Place entire assembly on south balcony or south roof. Or place it inside a solar greenhouse, where ambient temperature is higher and there is no snow. If the assembly is to be used in a very cold climate, install, along bottom of tank, an 80-W electric heater strip to be left on during the coldest 3 months (at total cost of $12). (Note: Danger of freeze–up is extremely small inasmuch as the equipment collects in the order of 3000 Btu on a sunless day in winter and the incoming cold water is much warmer than 32°F.) Install perforated near-horizontal baffles in tank to minimize mixing of hot and (input) cold water.

MAJOR MODIFICATIONS

Scheme S-165-rev.a

Replace the cylindrical tank with a very thin water–type collector—one that has very small thermal capacity. Make the insulating canopy correspondingly smaller. Employ a storage tank in the basement, and, when the sun shines, circulate water from this tank to the collector in any conventional way.

Incorporate in the collector some low-power electrical heater strips which, on cold nights, supply enough heat so that the collector does not cool down below 40°F. Then no antifreeze is needed, and no heat exchanger.

DISCUSSION

The crux of this scheme is to use a collector assembly that is insulated so well (with foam-type insulation on sides and top and a stable, thermally stratified air-film below) that the heat-loss on a cold night is so low that even a small amount of electrical heating will prevent freeze-up. Thus plain water may be used, no antifreeze is needed, and drain-down is not required. Essential to the scheme is the all-around insulation, and essential to this is (a) the inverted geometry—radiation striking the underside of the collector proper, and (b) large ratio of gross radiation-accepting aperture to the "hot aperture," that is, a moderate amount of concentration so that the area of the absorber proper is relative small.

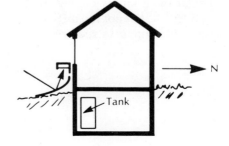

Scheme S–165–rev.b.

As above, except place the collection system just above, and north of, the north eaves of the house. To support the device here, the occupant can construct a framework that may be partially disguised as a portico or pergola.

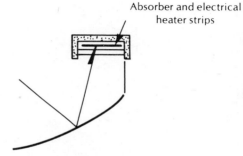

Absorber and electrical heater strips

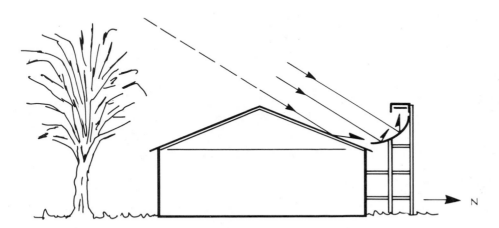

Vertical cross section, looking west. Note that no shrubs, and no trees less than 30 ft. tall, can shade the collector.

Notice how ideal this site is for a collector, at least as far as receipt of radiation is concerned; being high up and on the north side of the house, it is well above all shrubs and, more important, there are no trees nearby to the south. If the roof becomes covered with snow, this snow reflects additional radiation toward the collector. The collector in no way encumbers, or shades, the gardens or patio at the south side of the house and in no way interferes with receipt of solar radiation by the big south windows. Access to the collector is difficult, but not too difficult if a ladder is incorporated in the pergola–like support structure.

SOLAR DOMESTIC-HOT-WATER SYSTEM EMPLOYING A COMBINATION COLLECTOR-AND-STORAGE-SYSTEM AND A COMBINATION REFLECTING-AND-INSULATING-MATTRESS

Scheme S-162
6/28/77

PROPOSED SCHEME

The heart of the system is a water-filled, black, 1-ft.-diameter, 10-ft.-long, cylindrical tank of galvanized steel. It is embraced (flanked) by a crude, cylindrical parabolic reflector of shiny aluminum which provides a concentration factor of 3. The system aperture is covered with a sheet of Kalwall Sun-Lite. The north portion of the tank is permanently insulated. The lower part of the reflector consists of a mattress comprising a flexible sheet of shiny aluminum backed by 4 in. of flexible foam insulation. On sunny days the mattress lies in its normal position on a base frame that holds it in roughly the desired shape, i.e., part of a parabola. At the end of the day the operator pulls a string which pulls the south edge of the mattress up and over the tank, insulation all the otherwise uninsulated portions of it.

The next morning the string is released and the mattress falls to its normal position. Taped along the underside of the tank bottom is a 100-watt electrical heater strip that insures that the tank can never cool down as far as 32°F. Every few weeks the owner may, if he wishes, adjust the tilt of the entire assembly to compensate for the changing altitude of the sun. The entire assembly can sit on a lawn or on a roof, or can be built into the south side of the house.

Note low center of gravity of tank, absence of antifreeze, and absence of a heat exchanger. No basement space is used. No controls (except the string, operation of which could, of course, be made automatic, at some extra expense). The whole assembly can be built at a factory and delivered fully assembled and ready for use by one truck.

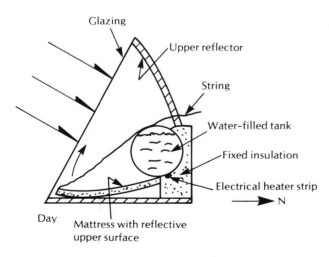

Day

Glazing
Upper reflector
String
Water-filled tank
Fixed insulation
Electrical heater strip
→ N
Mattress with reflective upper surface

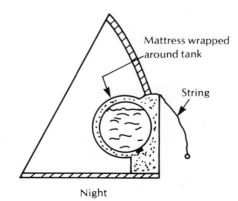

Night

Mattress wrapped around tank
String

224

MODIFICATIONS

John Golder's Scheme

Here, the lower (south) portion of the reflector is rigid and fixed, and the *upper* portion is flexible. At the end of the day this portion is wrapped around the tank to insulate it. At all times the cylindrical tank is enclosed in a loose-fitting jacket of Tedlar or Kalwall Sun-Lite; the air between jacket and tank helps reduce heat-loss. The scheme is described in the *Santa Cruz Alternative Energy Newsletter* of August 1978.

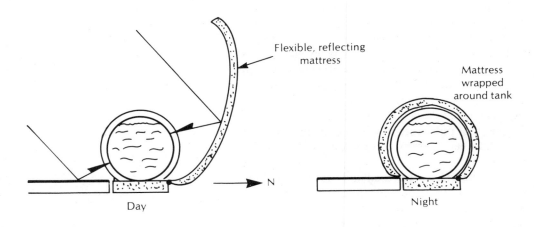

Flexible, reflecting mattress

Mattress wrapped around tank

Day

Night

Scheme S–162a

Instead of a flexible mattress, use two rigid insulating plates that are attached by hinges and can be swung so as to form an enclosing insulating box. This scheme is much like the one described on p. 64 of S.C. Baer's book (1975) *Sunspots*.

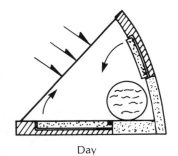 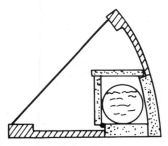

Day

Night

POISSON'S COLLECTOR BOX WITH TILTING BEAD–FILLED WINDOW

INTRODUCTION

This system, invented by Leandre Poisson of Solar Survival, Harrisville, NH, was displayed on June 24, 1977 at the Toward Tomorrow fair in Amherst, Mass. I examined it, operated it, and formed a high opinion of it. The following account is at least approximately correct.

DETAILS

A tank of water sits inside a truncated box that has reflective, insulating sides. The entire sloping south face of the box consists of a window that is glazed with two sheets of Kalwall Sun–Lite, which are 3 or 4 inches apart. During a sunny day, the space between the glazing sheets contains only air, and solar radiation passes through and is absorbed by the tank. At the end of the day, the operator opens a tiny gate near the upper edge of the window and this releases a large quantity of styrofoam beads which promptly fill the space between glazing sheets and complete the insulation of the box. At the start of the next sunny day the operator swings the lower edge of the window far upward so that all of the beads fall downward into the storage cylinder integral with the north edge of window, and he then closes the gate and lowers the window.

Perspective view

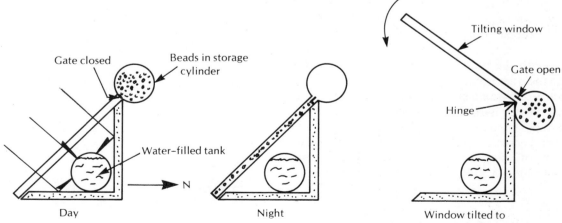

Day

Gate closed

Beads in storage cylinder

Water–filled tank

N

Night

Tilting window

Gate open

Hinge

Window tilted to sequester the beads

Scheme C-73½
11/6/78

SOLAR DOMESTIC-HOT-WATER SYSTEM EMPLOYING AN INDOOR ROLL-UP COLLECTOR MADE OF SYNTHETIC RUBBER

PROPOSED SCHEME

Inside a south-facing room, close to the room's large window-wall, there is a 6-ft.-long, 12-in.-diameter horizontal roll—a rolled-up, water-type collector housed in an insulating box that rests on the floor.

Whenever, during the daytime, the room is cold, the occupant leaves the roll undisturbed, i.e., leaves the window area clear so that solar radiation can penetrate deep into the room and warm it. But when, on a sunny day, the room is warm enough, the occupant deploys the collector. Using a rope-and-pulley system, he hoists a horizontal support bar which unfurls (hoists) the collector and exposes its 8-ft. x 6-ft. area to solar radiation. Plain water (potable water—no antifreeze, no inhibitor) from the 80-gal., domestic-hot-water tank in the basement is circulated through the collector by means of a small centrifugal pump.

The collector consists of three layers, all of which are flexible: (a) a thin layer of plastic, (b) a serpentine array of rubber absorber mat, discussed below, and (c) a 1-inch-thick, insulating, quilt-like backing. The flexible plastic serves as informal glazing for the collector. (The south window-wall has, of course, its own rigid double glazing.)

The heart of the collector is a black synthetic rubber absorber mat such as is produced by Bio-Energy Systems, Inc., of Spring Glen, NY, and is described on p. 109. it is an extruded assembly (EPDM, ethylene propylene diene monomer) that is 4.4 in. wide and about 150 ft. long. It includes six parallel, 3/16-in.-ID tubes, 0.7 in. apart on centers, joined by a thin web. It withstands high temperatures, can be bent or folded sharply, and is unaffected if the water in the tubes freezes. (Because the system is indoors, it is unlikely to cool down to 32°F.)

Because the coolant is plain water, no heat exchanger is needed. The flexible rubber tubes extend all the way into the basement, close to the 80-gal. tank. The headers are situated here. Connecting the tubes to the headers is a task that can be done quickly by hand; no tools are needed.

Being in a warm environment (indoors), the collector can operate efficiently, even if its output temperature is as high as 150°F. Heat that escapes from the collector helps heat the room.

At the end of the day the occupant presses the collector against the window by means of one or two pressure bars. Thus, at night the collector, which includes a quilt, acts as a thermal curtain.

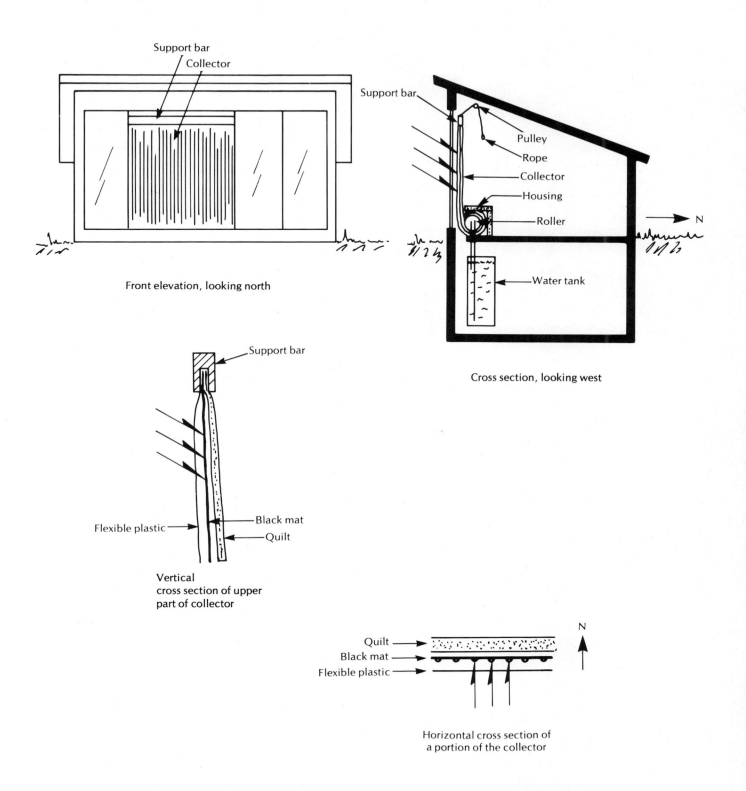

227

Support bar
Collector

Front elevation, looking north

Support bar
Pulley
Rope
Collector
Housing
Roller
Water tank

Cross section, looking west

Support bar

Flexible plastic
Black mat
Quilt

Vertical
cross section of upper
part of collector

Quilt
Black mat
Flexible plastic

N

Horizontal cross section of
a portion of the collector

MODIFICATIONS

Scheme C–73½ a

Make the roller and the tank one and the same. Obtain a 6-ft.-long, 1-ft.-diameter cylindrical horizontal tank and use it as a roller. Because, being filled with water, it is heavy, ball bearings should be used. When the collector is being rolled up or unrolled, the roller (tank) turns approximately two revolutions; the flexible tubes of the ends of the mat, and likewise the flexible tubes that serve the tank, can accommodate this.

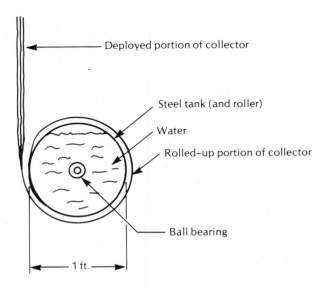

Deployed portion of collector

Steel tank (and roller)

Water

Rolled-up portion of collector

Ball bearing

1 ft.

PART 8

Greenhouses

INTRODUCTION

Many persons, or teams, have designed and built greenhouses that, even in winter, are almost 100% solar heated. Many such projects have been very successful.

A variety of solar greenhouses have been described by R. Fisher and B. Yanda in their book *Solar Greenhouses*, published in 1976 by John Muir Press, Sante Fe, NM 87501, $6. An especially interesting solar greenhouse, employing a water-spray type of collector, has been built by D. R. Hendricks *et al.* of the Hayes Regional Arboretum in Richmond, Indiana; it is described in detail in my book *Solar Heated Buildings of North America: 120 Outstanding Examples.*

On November 19–20, 1977, a two-day conference on solar greenhouses was held at Marlboro College in Marlboro, VT; the 284-page *Proceedings,* priced at $9., was published in March 1978.

Vast numbers of solar greenhouses would be built, presumably, if ways could be found of reducing heat-loss on cold nights without greatly increasing the cost of construction. Several schemes that might be effective are proposed below.

NEAR-100%-SOLAR-HEATED GREENHOUSE EMPLOYING DUAL-FUNCTION TRICKLE-WALL

SUMMARY

The proposed near-100%-solar-heated, low-cost greenhouse includes the following features:

- Very large (27-ft.-high) radiation-receiving aperture, double glazed

- Optimum apportioning of received solar radiation to (a) direct heating (25%) and active collection for storage (75%)

- Built-on-site, cheap, durable, trickle-type collector, or *trickle-wall*

- Situating the trickle-wall so deep within the main room that (a) heat losses from the trickle-wall constitute *gains* for the room and (b) the ambient temperature in front of the trickle-wall is so high that collection efficiency is especially high

- Making the trickle-wall do double duty: on sunny days it collects energy, and on cold nights it dispenses energy to the room

- At night the ground-level room is fully insulated

A modified scheme (S-144a), employing a truncated (cheaper) collector, may be suitable for use in warmer locations.

PROPOSED SCHEME

The south wall of the greenhouse is double glazed with plastic and slopes 60° from the horizontal. The north wall (*trickle-wall*) slopes 85° from the vertical—it has an upward, south slope. It includes two nesting sheets of corrugated aluminum with the corrugations running up and down. The south (under) surface of the south sheet has a non-selective black coating. (Note: *One* corrugated aluminum sheet might suffice; the north sheet could be replaced by say, a sheet of cheap, thin, waterproof plastic.) Water trickles down between the two sheets, making closest contact with the south sheet. The water is fed from many 1/16-in.-dia. holes in an upper horizontal header-pipe, which extends the full length of the greenhouse. The water is collected at the bottom of the trickle-wall and runs into an insulated tank situated within the earthen base of the greenhouse, i.e., under the earth floor. The tank consists of a long section of 4-ft.-dia. corrugated metal culvert, with plastic liner. The north side of the trickle-wall is well insulated. The overall height of the greenhouse is 25 ft.; the width of the earthen base (region for growing plants) is 15 ft.; the height of the solar-radiation aperture is 27 ft. About 12 ft. above ground level there is an insulated *divider-floor*, with an aluminum-foil-covered upper surface, which slopes upward-to-the-south at 25° from the horizontal. Close beneath it, on sunny day, there is an insulated cover, hinged along the south edge, that, at the end of the day, is swung down so as to lie close against, and insulate, 70% of the glazed area between ground and divider-floor. Affixed to the lower portion of the cover is a junior cover, that can be used to insulate the remaining (lowest) portion of the glazing. No insulation is provided for the attic glazing area. There is a second, or alternative, feeder pipe (the lower header-pipe); it is situated near the north edge of the divider-floor; it can be used at night to distribute hot water (from the tank) to the lower portion of the trickle-wall.

Details of the hinged insulating cover are shown in the accompanying sketches. Spring clips for holding the insulating members firmly against the glazing are not shown.

The building *ends* are opaque and heavily insulated. The faces that are toward the interior of the building are aluminized. The earthen floor is largely covered with plants 1 to 3 ft. high.

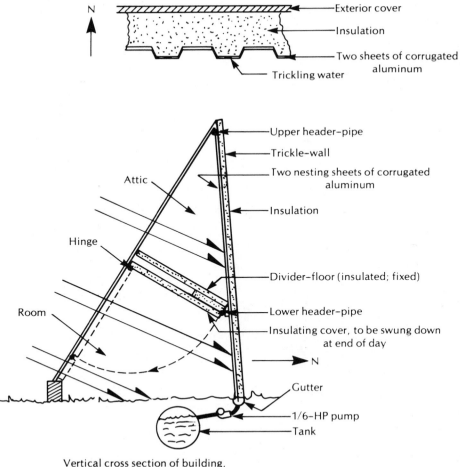

N

Exterior cover

Insulation

Two sheets of corrugated aluminum

Trickling water

Horizontal cross section of portion of trickle-wall

Upper header-pipe

Trickle-wall

Two nesting sheets of corrugated aluminum

Insulation

Attic

Hinge

Divider-floor (insulated; fixed)

Room

Lower header-pipe

Insulating cover, to be swung down at end of day

N

Gutter

1/6-HP pump

Tank

Vertical cross section of building, looking west. Condition: sunny noon

232

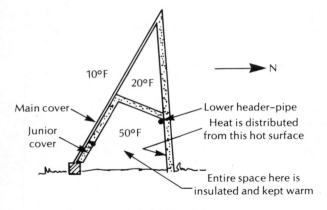

Arrangement at night in winter

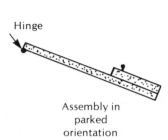

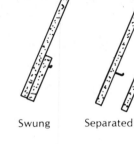

Diagrams showing the main cover and junior
cover in four states of deployment

OPERATION

Winter

During a sunny day in winter all parts of the trickle-wall receive
solar radiation and all parts are supplied with water via the upper
header-pipe. Much energy is carried by the trickling water to the
tank.

   During the night, the lower half of the building (the room)
loses only a little heat because of the all-around insulation. Losses
are made up by heat from the lower half of the trickle-wall: this
part of the wall is fed (via the lower header-pipe) with hot water
from the tank whenever the temperature in this room threatens to
drop below 50°F. The distribution of heat is efficient because of
the large area of radiating surface (trickle-wall) and because the
black coating is of *non*-selective type. No attempt is made to keep
the attic warm at night.

Summer

During a hot day in summer the divider-floor provides much shade. Also, various vents (not shown in sketches) are opened. If additional shading is needed, the movable insulating cover may be swung down to provide it. No attempt is made to shade the attic, but the attic is vented.

Percent solar heated: Almost 100% (a guess).

DISCUSSION

The amount of solar radiation received in mid-winter is very large, inasmuch as the glazed aperture is large (27 ft. high) and slopes steeply. At noon on a sunny day in winter, 75% (a guess) of the direct radiation reaches the trickle-wall and the rest strikes the earthen base itself. This constitutes, I guess, a near ideal apportionment. At times a few hours before or after noon, the fraction reaching the trickle-wall is greater.

The collector is cheap, being built on-site with cheap, standard, 24-ft.-long sheets of corrugated sheet aluminum. There are few pipes, few connections, no heat exchanger, no antifreeze. The collector should work well hydraulically and thermally despite the fact that it is nearly vertical. Because the two corrugated sheets nest fairly closely together, the descending water descends with some friction and turbulence, insuring good contact with the under sheet—the south, radiation-receiving sheet. Thus, the efficiency of heat transfer is high.

Collection efficiency should be especially high because the collector is located deep inside the building, i.e., it is in contact with air that is warm (60 to 80°F. typically). In any event, heat losses from the trickle-wall constitute heat *gains* for the room. The overall collection efficiency of the system as a whole is about 75% (a guess) even on overcast days or days with clouds coming and going. The start-up time is practically zero.

If some moisture escapes from the trickle-wall into the room, this may be positively helpful: plants like moisture.

The triangular shape of the building cross section provides mechanical strength and makes construction easy. Component walls can be constructed flat on the ground, then swung up into final position.

The divider-floor and parked movable insulating cover are "edge-on" with respect to the sun's rays at typical time of day in winter. No significant obscuration is produced.

When the movable cover swings down, its lower edge is at all times at least 3 ft. above ground level. Thus it does not strike modest-height plants.

Operation of the movable cover could be partly or fully automated.

Spring clips, latches, etc., could be provided for holding the cover (or its edges, at least) tight against the glazing. Such tight fit would pay off on very cold nights.

Because all components of the general enclosure (other than the earthen base) are of thin, light-weight materials, the thermal capacity of such components is small and the warm-up time of the system is short.

If, on a sunny morning (say), the operator wishes to warm up the room as fast as possible, he merely turns off the pump that sends water to the trickle-wall. Practically all of the solar energy will then go into heating the room.

Winter nighttime heat-loss from the lower half of the building is very small, inasmuch as there is good all-around insulation of this lower half. Loss from the upper half—the attic—is irrelevant, this half being thermally divorced from the lower half and being permitted to cool down indefinitely at night.

If the storage tank were to leak slightly, this would do little harm: the water that leaks out would help the plants to grow. Leakage of heat, likewise, is not very harmful: the heat that leaks out helps keep the room warm. The warm earth contributes to thermal storage.

Carrythrough is very long because (a) much heat is stored in the tank and the earth, (b) the lower half of the building is double glazed and, at night, has all-around insulation, (c) the lower half may be allowed to cool down to 50°F at night (thus losses are small), and (d) no attempt is made to heat the attic.

No separate heat distribution system is required—the lower half of the trickle-wall serves this purpose. It is used at night to transfer heat from the tank to the lower half of the building. Much of this heat is transferred by radiation; some is transferred via conduction and convection. The amount of heat transferred by radiation is large because the black coating is non-selective.

The big movable insulating cover is "self-stored" when swung up and out of the way in the morning.

The auxiliary heat source, i.e., oil heater or wood stove, if such is needed, can merely deliver its heat to the tank. A very small auxiliary source suffices inasmuch as it can be operated around-the-clock to provide energy needed during limited periods only.

The overall system makes it easy and cheap to water the plants with warm water, i.e., water from the tank.

In summer, ample shading is provided by the divider-floor, augmented, when necessary, by the movable cover. Large vents help keep the room from becoming too hot.

MODIFICATIONS

Scheme S-144a for Warmer Climates

Here the glazing is single. The attic is eliminated. One header-pipe suffices. The junior cover of the movable cover system can be omitted.

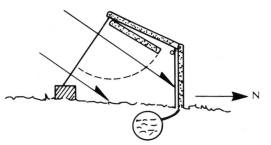

SOLAR GREENHOUSE EMPLOYING HINGED, THREE-PURPOSE PLATE

SUMMARY

The unique feature of the design is a large, hinged plate that serves three purposes. On sunny days it serves as a water-type collector. On cold nights it is swung down and insulates the plant-growing region, and in addition it serves as a hot-water-type radiator to dispense heat to this region. The water-filled thermal storage tank is situated in the earth beneath the plants.

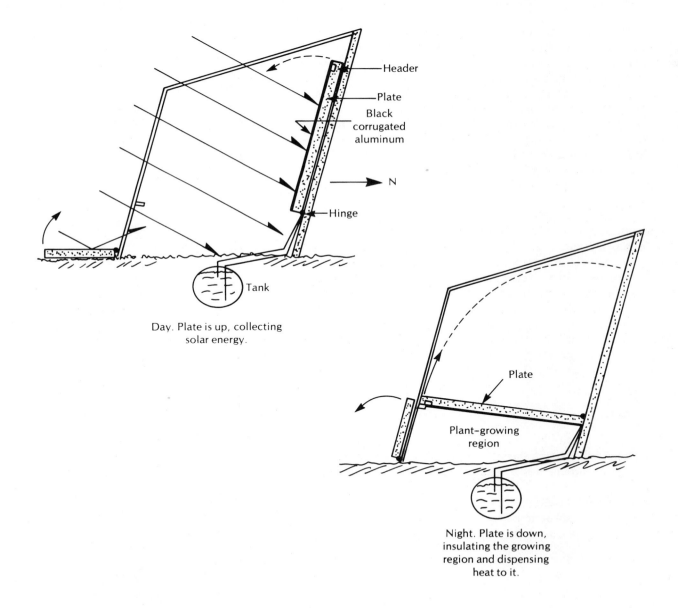

Header
Plate
Black corrugated aluminum
N
Hinge
Tank

Day. Plate is up, collecting solar energy.

Plate
Plant-growing region

Night. Plate is down, insulating the growing region and dispensing heat to it.

PROPOSED SCHEME

The south wall and roof of this New England solar greenhouse are double glazed. The north wall is heavily insulated. Both walls and the roof slope upward to the north as indicated in the sketches. Of central importance is a large multi-purpose plate that, by day, lies close against the north wall; it is attached to the wall by means of hinges that are 4 ft. above ground level. On sunny days much radiation strikes the plate and is absorbed by its south face, which consists of a sheet of corrugated aluminum that has a non-selective-black coating. Nested close against this sheet there is a similar sheet, and water trickles down between them, picking up heat and carrying it to the storage tank. The trickling water is dispensed to the plate by a header pipe served by a 1/6-HP centrifugal pump. The supply and return pipes pass close to the hinges and, here, are flexible.

The hot water from the collector flows to a large, waterproof, insulated tank buried a few feet in the earth beneath the region in which plants grow.

At the end of a sunny day in winter the plate is swung downward so as to form an insulating cover for the plant-growing region. If the temperature in that region falls below 50°F, hot water from the tank is circulated to the pair of corrugated aluminum sheets (now constituting the underside of the plate), and these dispense heat (by convection and radiation) to the growing region. The same pump, header, supply and return lines mentioned above are used.

The lower portion of the window area is covered at night by means of a small insulating panel which by day lies flat on the ground outdoors and reflects solar radiation into the greenhouse.

Manual operation of the plate is simple because counterweights are provided.

The headroom beneath the "down" plate is minimal. It could be increased, but only if the greenhouse were made narrower or taller.

The system has many good features. Most of these are identical to those listed for Scheme S-144, described on previous pages.

MODIFICATIONS

Scheme S-146a

Make the upper south portion of the building curved; employ ribs; employ two plastic glazing sheets held apart by air pressure. Also, add a counter-sloping north wall which would provide added strength and provide space for storing tools and supplies. (J. C. Gray has suggested making the main north wall vertical, which would simplify construction and improve the appearance. The amount of solar radiation received would be slightly reduced.)

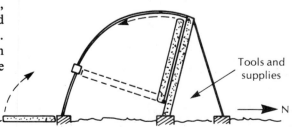

Tools and supplies

N

PART 9

Comedy Relief

INTRODUCTION

People tend to take solar heating very seriously.

They should.

I do too.

But sometimes we have to relax and laugh at ourselves, at our bumbling government agencies, and at our pontificating colleagues. (Do we ourselves ever pontificate? Ridiculous question.)

Presented below are several essays that are partly serious, partly intended to cause tense readers to relax.

Every lone-wolf inventor (but no big-corporation executive) knows that when the US Government hands over $500,000 to a big corporation to develop a reliable and cheap solar heating system, the corporation struggles valiantly and produces a system that is hopelessly complicated and expensive. Contrariwise, if a lone-wolf inventor, using his own savings, starts to design and build a solar heating system for his own house, he has a good chance of ending up with a system that does the job and costs little.

Bewildered by this paradox, I consulted economists, engineers, and corporation heads. But to no avail.

Finally, I consulted the famous Parkinson—the man who discovered that, nine times out of ten, when you enter a crowded post office and carefully choose the shortest line to wait in, *any other choice* would have been preferable.

Parkinson listened to my question, gazed at the ceiling for a minute or two, and said:

The explanation is simple enough. When you pay a corporation $500,000 to invent a solar heating system, the corporation officials have no choice at all. There is no way that they can escape making a complicated, much-too-expensive system. They are 100% trapped by the Principle of Pyramiding Pride.

They order all of their department heads to cooperate fully. Each department heads enlists the aid of his best engineers. Each engineer tries to design the very best equipment.

The public relations director calls a press conference and explains that the company is about to score the long-needed breakthrough in solar heating, is about to Lead the Way out of the energy crisis.

Hearing this, the engineers redouble their efforts to devise the very best equipment, so that everyone will agree that the government has gotten its money's worth—government officials will be wreathed in smiles, the technical journals will publish salvoes of praise, and the company stockholders will cherish feelings of real reverence for their company.

The project is completed on schedule. The result is a technological masterpiece. A joy to behold. It is praised by everyone—government officials, company heads, press, and public. (The *user* soon finds the equipment to be too complicated and costly to maintain, and (quietly, shamefully) he abandons it.)

I was shocked. "This can't be true," I pleaded. "Surely the corporation heads are intelligent. They know the meanings of the words *reliable* and *cheap*."

"But they have no choice. Parkinson replied. "Surely you remember the Central Africa writing-machine competition?"

I did not, and he explained:

Fifteen years ago the United Nations awarded identical contracts to two corporations: Trans–World–Products and the Sam Botts Co. Each was given $1,000,000 and told to design a writing machine that would be truly suited to African countries. The device was to be capable of writing in small letters or large, in English, French, German, or Swahili. It was to withstand tropical dampness and floods.

The Trans–World–Products engineers went to work with a will. They used up all the money and time allowed. They produced a 200–lb. stainless-steel machine, housed in a fiberglass container, which included a rechargeable battery, a 5–year desiccant cartridge, flotation gear, and a 100–page maintenance manual written in twelve languages. Although the first model cost over $100,000 to build, later units could be mass-produced, it was claimed, for only $1500. The device was a marvel to behold, and the world was lavish with its praise. The president of the company was given a 15% salary increase, the department heads were given bigger offices. Even the stockholders in the company felt ennobled by being involved in such a successful and altruistic project. TWP's final report (in four volumes and weighing 8 lb.) is available in all major libraries.

The Sam Botts Co. took no visible action for many months. Old Man Botts said nothing to his department heads. He asked no one for help. He built nothing. Day after day he sat in his small office staring off into space. Finally, he mailed off a small package (a Manila envelope, which required 30¢ postage) to the sponsoring agency. The envelop contained an ordinary Faber Co. wooden pencil, a check for $990,000, and a brief note which read: "This machine (pencil) meets the requirements—it writes in any language, is unaffected by damp climates, and, when caught in a flood, floats. Am returning the money we didn't need. Yours truly, S. Botts."

The sponsoring agency was furious with Botts. The press ridiculed him. The stockholders felt crushed; they cut his salary and eventually eased him out of the company entirely.

Today, there are 3,237,000,000 wooden pencils in use in Africa. No second TWP machine was ever built.

Before I could catch my breath, Parkinson said, "Sorry. Must leave. Am to attend the dedication of the new NSF–funded solar heating system of the local high school. They say it is a technological masterpiece."

SEARCH FOR A SITE FOR SERI: WHAT IS THE SITUATION NOW, IN 1984?

(This essay was written in mid–1976 when DOE was continuing to dawdle in making the choice of site for SERI.)

Today—mid–1984—it is hard to believe that our government's search for a site for SERI (Solar Energy Research Institute) is in its tenth year, with no solution in sight. How simple it seemed, back in 1974, to pick a good location for this national institute that was to solve our energy problems.

What an agonizing decade it has been! The public's first awakening to energy shortages came in 1973, with the Arab's four-month embargo of oil. Motorists had trouble buying gasoline. Homeowners worried that their furnaces might run out of oil. During the second embargo (6 months, in 1977), automobile traffic came almost to a halt, and millions of home-owners learned how to keep their furnaces going with logs (cut, usually, from neighbor's shade trees). The third embargo, in 1979, was even more serious; few persons—other than politicians, TV celebrities, and professional athletes—were allowed to use cars. Cruising policemen cruised on bicycles. General Motors shut down its production of Cadillacs and started production of its now-famous Hydromatic roller-skates. Homeowners, having no more wood to burn, took to burning mail-order catalogs, government reports, and telephone directories (AT&T boasted of its role in "helping keep America warm").

But why, with the need for exploiting solar energy so obvious, did our government continue to delay in choosing a site for SERI? Perhaps the best answer is that given by my colleague Parkinson in his recent pamphlet "Is This Delay Neces–SERI?" At the recent meeting of our Chamber of Commerce, he summarized his views more briefly, thus:

The basic cause of the delay was the sheer urgency. To accelerate the program, Congress repeatedly voted SERI larger and larger sums. In 1976, the budgeted figure was $5 million. In 1977 it was $27 million. In 1978, $128 million. By 1980, it was $990 million.

Each time the figure was increased, a new set of proposals was called for. Each bidder strained to upgrade his bid, to make it more glamorous than his competitor's. The total number of groups bidding (states, regions, cities, universities, conglomerates) increased rapidly, soon exceeding 1000. Several bids came from Canada, one from Japan, one from Australia (its bid was entitled "Let SERI Waltz with Matilda"). Even Arabia eventually submitted a bid (with the apology "Better late than oily").

Our government tried hard to be fair. Each time it called for new bids (and issued a new "clarification of requirements") it allowed the bidders six months to respond. It arranged mammoth meetings to clarify the clarifications. An ambitious firm in Baltimore started a weekly newsletter (at $500 per year) solely devoted to advising on the best way of writing proposals for SERI; the newsletter was called *SERI–Cerebration Weekly*.

Also, the government repeatedly upgraded its choice-making process. In 1976, the final choice was to have been made by a deputy director of ERDA (the agency later known as the Endless Reconsideration and Deliberation Agency). Late in 1977, the agency head announced that he personally would make the decision. In mid–1978, the President declared that the National Security Council would resolve the issue. Days later, in an unprecedented joint session of House and Senate, the Congress voted 689 to 0 to create a "Joint SERI Committee" to make the fateful choice. There were rumors that the Supreme Court might step in.

At this point Parkinson was interrupted by an angry voice from the rear of the hall, "But why did the delay go on and on? Why couldn't the government view the facts and make the choice?"

Parkinson paused, seemed lost in thought for a few moments, adjusted the whale-oil lamp that illuminated the lectern, and replied:

Because the facts were changing so fast. The competing groups kept changing and improving their proposals. Each group, having already invested several million dollars in writing, illustrating, and printing its various proposals, strained itself to the utmost to make its proposal even more impressive.

Take 1977. Colorado had announced the availability of 2 square miles of choice land at reasonable cost. California offered a 4-square-mile tract free. Arizona offered a 5-square-mile site complete with an abandoned hotel, a small airport, and an outdoor movie theatre that had a curved screen half the size of the great French solar furnace. New Mexico, besides offering a 10-square-mile tract in a region that received rain only once or twice a year, boldly named the top 10 men who would head up the institute. New England offered a choice of 37 alternative sites, listed the 100 top administrators to be employed, and identified 20 specific projects to be undertaken.

But by 1978 the competing proposals had gotten out of control. Colorado described 200 proposed projects, assigned detailed priorities, and presented estimates of times-of-completion of the 10 principal projects. Arizona promptly countered with a 350-pound proposal that included estimates of the dates of start of mass production of the 50 main products developed. New Mexico was the first to list the expected profits from the mass-produced products and the expected amounts the government would obtain in taxes.

In 1979 the competition centered on the size and quality of staff. Colorado proudly announced it had tentatively signed up more than 100 expert consultants from universities and industry; three Nobel prize winners were included. Within two weeks Arizona countered with a list of 500 consultants, including 75 listed in *Who's Who* and 350 carrying American Express or Gulf Oil credit cards. New Mexico's revised proposal listed 1200 registered professional engineers, two retired Supreme Court justices, two alleged sons of Howard Hughes (and three other listed in *Who's Hughes?*). The New England group, buying prime time on a nationwide TV hookup, announced that its list of consultants included every college president in New England, the entire faculty of MIT, and the heads of all Rotary and Kiwanis clubs. It said also that it had arranged to secure the help of the entire membership of the National Academy of Sciences (for the nominal fee of $1 the first day, $2 the second day, $4 the third day, etc.)

So far, the competition had been clean. But now Florida created a sensation by passing a law that automatically made every man, woman, and child in the state a consultant. Also, it promised that, should SERI—established in Florida—fail to lead this country to energy self-sufficiency by 1990, Florida would, as forfeit, donate the entire 1990 orange crop to the US Treasury. The responses of other states were even more offensive but are so fresh in your minds that I need not review them here.

But let me close on an optimistic note. During all those hectic years of pyramiding proposals, inventors everywhere were quietly working away, long forgotten by the professional (career-men) proposal writers. In 1982 the low-cost photovoltaic cell was perfected and went into mass production. The one millionth solar heated house was built. The 56th and 57th solar towers were completed (they consist of the Golden Gate Bridge towers, each served by 100 pontoon-mounted tracking mirrors), and New York's World Trade Center towers are now being converted. Windmills have taken over almost half of the load of Con Ed, and nearly every rural town has an animal-waste methane plant in operation. A special branch of the US Patent Office has been set up to process solar-energy inventions only. France's prestigious solar-and-wind conglomerate "Insol-Vent" has become so prosperous that it has proposed a merger with Detroit's General Motors and Manure Company.

As you know, a few states have become sick of the entire subject of SERI. Last week the Texas legislature declared that if SERI is assigned to Texas, Texas will quit the Union.

PARKINSON'S PARADOXES CONCERNING HEAT-LOSS THROUGH WINDOWS

By exploiting his powers of concentration to the utmost, Parkinson has arrived at the following important conclusions:

1. Adding a thermal shutter to a window of a typical house is a mistake. It *increases* the total heat-loss of the house. How so? Adding a shutter tends to keep the house a little warmer, and, other things being equal, the warmer a house is, the more heat it loses.

    Of course, adding a shutter to a given window reduces the loss through this particular window. But what counts is the house as a whole; and the addition of a shutter increases the losses through all the other windows and through all the walls.
    Moral: Remove all shutters so that the house will be cooler and will lose less heat.

2. A great way to reduce the heat-loss through a double-glazed window (see the first accompanying figure) is to use a small blower to suck air into the house via the space between the two glazing sheets. Just make a small slot at the bottom of the outer sheet and another slot at the top of the inner sheet. The stream of incoming fresh air will pick up the heat that passes through the inner sheet and carry it back into the house. (See the second figure.)

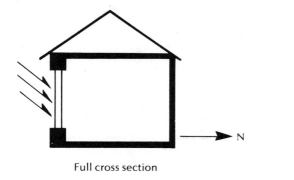

Full cross section

House with ordinary, double-glazed window.

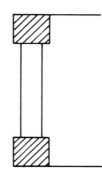

Detail of double-glazed window

Of course, if air flows into the house in this manner, an equal amount of air must be allowed to flow out of the house. This will entail some loss of heat, unless one confines the outgoing air to a tube and bends the tube around so that the air which it discharges flows directly into the slot in the outer glazing. (See the third figure.) A small residual heat-loss is avoided if the tube, instead of being partly exposed to outdoor air, is kept entirely within the house. (See the fourth figure.) Unfortunately, a little heat is lost through the outer glazing layer, especially if there is turbulence in the air flowing in the space between glazing sheets. This loss can be reduced by reducing the airspeed here, i.e., by cutting down on the power supplied to the blower. Best results are obtained with the blower shut off entirely and the two slots closed off. (See the fifth figure.)

To appreciate the overall, four-step, improvement made, one may compare the fifth figure with the first.

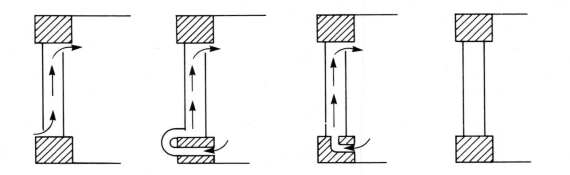

LETTER CRITICIZING THE FORD FOUNDATION FOR OVERLOOKING AN ENERGY SOURCE THAT SOME PEOPLE THINK IS IMPORTANT

Dr. David Freeman, Director   April 10, 1974
Energy Policy Project
Ford Foundation
PO Box 23212, Washington, D.C. 20024

Dear Dr. Freeman:

With alarm I notice that your new, fine, 82-p. report "Exploring Energy Choices" makes no mention of our major source of energy, namely the source that, in USA alone:

Grows $100 billion dollars worth of wheat, corn, etc., each year

Grows 5 trillion trees and shrubs, 17 trillion flowers

Provides daytime illumination for 200 million people

Melts 10 trillion tons of snow and ice each spring

Vaporizes 6 trillion tons of water that results in rainfall supplying 3 million acres of farmland, 1000 rivers

Warms 5 trillion tons of coastal water in which 10 million people swim

Produces winds that drive 100,000 windmills, propel 700,000 sailboats, lift 200,000 kites, dry 1 billion freshly laundered sheets, purge smelly air from 1 million city streets

Provides a universally visible time-piece by which people can distinquish morning, noon, and night (except in Los Angeles)

Sincerely,

William A. Shurcliff
Solar Heating Consultant

PART 10

Miscellaneous Essays

Here I include miscellaneous essays on several subjects that tie in closely with the subject of solar–heating–system inventions.

COLLECTION EFFICIENCY: THE WRONG CRITERION

Most persons who sell solar collectors talk a lot about collection efficiency. They proudly claim that their product has, say, an efficiency of 65%.

This sounds interesting, but it is not.

Why not? For many reasons—technical reasons and strategic reasons.

TECHNICAL REASONS

1. Efficiency of a given collector can vary enormously, depending on many circumstances, such as direction and intensity of radiation, outdoor temperature, windspeed, and storage tank temperature.

2. The efficiency can be as high as 1.00. (It can be even higher if the outdoor air happens to be hotter than the storage tank.)

3. The efficiency can be as low as 0.00. (It can be even lower—it can be negative—if the outdoor air is very cold and the storage tank is very hot.)

4. There is no way of specifying a meaningful set of "typical conditions." Typical for what part of the country? What time of year? What extent of cloudiness? What size of storage system? What setting of the room thermostat?

STRATEGIC REASONS

1. What the buyer really wants is a collector that collects much heat at low cost. It is the ratio of *heat collected* to *overall cost* that counts. (Efficiency just is not relevant. If the most efficient collector is the most expensive one, it may be the very worst buy. As an analogy, consider the speed of a sailboat: When you are choosing which sailboat to buy, you probably give much attention to seaworthiness, stability, ease of handling, durability, and cost—and little attention to speed. Other things being equal, speed is an asset. But other things are not equal and the fastest boat is likely to rank low in many important respects.)

2. The other main criteria are durability, reliability, and ease of repair. (Efficiency, I repeat, is irrelevant.)

3. The fact that most of the efficiency claims are practically identical—around 65%—strongly suggests that this whole subject is a red herring. How much pleasanter it is for the dealers to talk about efficiency than about the ugly subjects of cost, durability, etc.

4. In general, efficiency of any process is of no interest unless there is a shortage of some sort. There are shortages of furnace oil and farm land, for example; therefore, it makes good sense to keep our furnaces working efficiently and keep our farm lands in productive use. But there is no shortage of sunlight; most of it is completely wasted. When you collect some of it, who cares how efficiently you collect it? (It is analogous to picking wild-flowers in Yellowstone Park: If I pick you a bouquet of such flowers, do you care how many flowers I overlooked? Do you care about my efficiency?)

5. The more the dealers talk about efficiency, the more our attention is distracted from what really counts: Btu's collected per dollar, durability, reliability, and ease of repair.

Of course, if you are comparing two kinds of collectors, and they are nearly identical as regards cost, durability, etc., it may be worthwhile to inquire about efficiency. I say "may" because the likelihood is that the efficiencies also will turn out to be practically identical.

Is there any reason for a person to become emotional and angry about the subject of efficiency? Yes, an excellent reason: Many of the innovative, truly low cost, solar-heating systems rank only *fair* or *low* on an efficiency scale—or simply have not been rated. Most of the passive systems have no efficiency ratings at all. If a dealer says, "Up with efficiency!" he is almost implying "Down with passive systems and down with active systems that are truly cost-effective."

IS IT MISLEADING TO DESCRIBE COLLECTOR PERFORMANCE BY MEANS OF GRAPHS OF EFFICIENCY VS. $[T_i - T_a]/I$?

SUMMARY

Yes, such graphs are misleading in several ways. They can give wrong ideas as to how a given collector will perform in various actual situations or how various different types of collectors will compare under a variety of real–life conditions. Also they distract attention from issues that are more important: cost and durability, for example.

INTRODUCTION

Many years ago solar collector designers employed various simple graphs of collector performance. For example, they prepared graphs in which collection efficiency was plotted against I (the level of irradiation) for a given value of T_i (collector–inlet temperature), T_a (ambient temperature), and coolant flowrate. They prepared many graphs for various different choices of these parameters.

But they were dissatisfied with such graphs because so many of them were needed and so many long test runs of the collectors—under a great variety of conditions—were needed.

More recently, collector designers have gravitated toward a short–cut method of presenting the performance data. First, they *correct* for T_a, the ambient temperature, by subtracting it from T_i, the collector–inlet temperature. Second, they normalize the level of irradiation, I, by dividing $(T_i - T_a)$ by it. In summary, the ordinate of the graph is efficiency and the abscissa is $(T_i - T_a)/I$. See the accompanying figure.

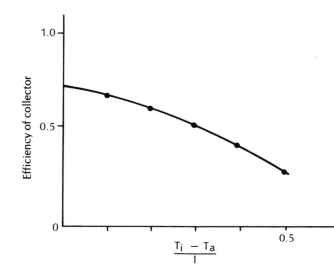

Deceptively straightforward–looking graph of the type that is in widespread use today

Thus such persons deal, mainly, with a single graph: a single curve. They assume it suffices. They assume it contains the distilled and consolidated information from a wide variety of tests. If they wish to compare two collectors, they obtain the two pertinent curves and stand back and compare them. Everything looks simple and scientific. (Very sophisticated designers know better, and they proceed more cautiously. My remarks apply to less sophisticated persons.)

WHY ARE SUCH GRAPHS MISLEADING?

Such graphs can be seriously mileading—not because of inaccuracy or dishonesty, but because of gaps in the underlying logic. In particular, they are misleading because so many important variables have been left out.

Here are some of the overlooked variables:

1. *Warm-up Time* When, on a cold day in winter, the sun suddenly comes out from behind a large formation of dark clouds, the collector warms up. But this takes time. For a typical water-type collector the warm-up time may be as long as 5 to 20 minutes. During the first minutes after the sun comes out, such a collector may be supplying no heat at all to the house. Temporarily the collection efficiency is zero. Have you ever seen a curve showing zero collection efficiency? Probably not. The persons who draw the curves usually shut their eyes to conditions giving zero efficiency.

    Is there some way such persons could make a standard allowance, or correction, for start-up time? No. Warm-up times vary greatly from one type of collector to another. (Passive systems start instantly: There is no delay at all.) Also, in some places, such as El Paso, Texas, the sky is almost always clear and warm-up time is not important; whereas, in Rochester, New York, heavy clouds come by frequently and warm-up time is very important. Also, the importance of warm-up time depends on the ambient temperature and on the temperature of the coolant delivered to the collector.

2. *Windspeed* Most collectors perform much better on calm days than on very windy days. Here again no simple correction can be made, because some kinds of collectors are much more affected by wind than others are. Collectors that are triple glazed or are vacuum jacketed are only slightly affected. Single glazed, poorly sealed collectors are greatly affected.

3. *Snow* Snow may quickly slide off a collector that is steeply sloping and is single glazed: Heat from within the collector may melt the underside of the quantity of snow, permitting easy slide-off. But if the collector is only gently sloping and is double glazed, slide-off may be delayed for hours or days depending on the outdoor temperature.

4. *Direction of sun* Collectors employing tubes that are well spaced and properly oriented may perform almost equally well

from 8:30 a.m. until 3:30 p.m. For flat-plate collectors this is far from true.

5. *Fraction of the incident radiation that is diffuse* Some collectors collect direct and diffuse radiation almost equally efficiently, but for others, especially collectors that have a high concentration ratio, this is far from true. Also, the extent to which a flat-plate collector collects diffuse radiation depends strongly on the tilt.

6. *Temperature rise of coolant in collector* The graph under discussion takes no account of coolant flowrate and no account of the extent to which the coolant heats up on passing through the collector. If its temperature is increased 100 degrees in Collector A but only 15 degrees in Collector B, this effect is ignored by the person who prepares the graph. He pays great attention to collector inlet temperature but none to the outlet temperature.

The accompanying figure shows in schematic manner how bewildering a graph may be if the experimenter goes out of his way to make tests under a great variety of extreme conditions, such as 30 seconds after the sun comes out from behind clouds, or high wind, or a heavy fall of snow some hours earlier, or sun at 60° from south, or radiation that is entirely diffuse, or a coolant temperature rise of 150 degrees. How can one draw a smooth curve through the plotted points? If, somehow, a smooth curve were drawn, of what use would it be?

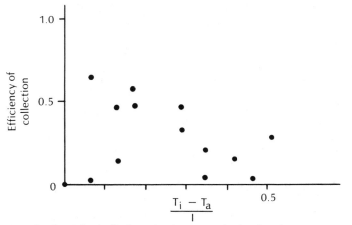

Here the (hypothetical) plotted points were obtained under a very wide variety of conditions. The scatter of the points reveals the shortcoming of such a graph.

POSSIBLE REBUTTAL

Someone might say: "The whole difficulty could be avoided by carefully standardizing the measurement process. We could make all measurements at about noon on a sunny day in March in El Paso, with baffles to block off the wind. We could give the collectors one hour to become thoroughly warmed up. With a little effort, we could obtain results that are simple and accurate, and give a simple smooth curve.''

This rebuttal is almost worthless. Such tests have little to do with real-life conditions of clouds coming and going, wind, snow, collector outlet temperature, etc. A collector that ranks high under the proposed standardized conditions might rank low under various real-life conditions.

Another type of rebuttal is this: Experienced solar engineers know the limitations of the graph and insist on obtaining much supplementary information. But (1) this robs the graph of much of its usefulness and (2) most people soon forget about the limitations and proceed to regard the graph as holy writ.

ADDED IRONY

The graphs are especially misleading when used to compare two wholly different types of collectors. Yet it is just here that designers and buyers are most in need of help.

Another bit of irony is that so many kinds of flat-plate collectors have almost identical curves, with the well-known and ludicrous result that each of dozens of manufacturers claims that its product has top, or near-top, ranking.

In summary, the graphs fail to provide a clear ranking of collectors of fairly similar type and can be especially misleading when applied to collectors of very different type. When needed most they are most misleading.

One main consequence of publication of such graphs is that the designers become bemused: They become convinced that they are proceeding in highly scientific manner. An aura of technical competence envelopes them. They forget that many important parameters have been swept under the rug and they forget that 5% differences in efficiency are trivial compared to the huge differences in cost, durability, etc.

The graph is a kind of placebo.

WARNINGS BY LUMSDAINE AND GOLDBERG

E. Lumsdaine of New Mexico State University, in a 1978 report ''On the Testing of Solar Collectors to Determine Thermal Performance,'' warned against the standard test procedure ASHRAE 93-77 on the grounds that the ''... curves can be invalid because thermal efficiency is influenced by such factors as collector tilt, angle, wind velocity, spectral distribution of radiation (percent of diffuse and direct) and the incident angle modifier as well as differences in the test fluid in the case of liquid collectors.'' He concluded that ''The magnitude of the errors can be quite large.''

M. H. Goldberg of Practical Solar Heat, Inc., in a report ''Solar System Performance and Component Reliability after 5 Years,'' sums the matter up: ''Collector designs, in the author's opinion, should be compared on the basis of actual heat delivered in actual service over a period of varied weather; otherwise false ranking of collector performance on basis of laboratory-condition testing is likely.''

Of course, the graph under consideration can be of some help—can be much better than no graph at all—if the user remembers that many important considerations have been ignored.

254

MARVIN SHAPIRO, YOU ARE WRONG! [WRONG ABOUT CRITERIA FOR JUDGING SOLAR HEATING SYSTEMS]

The dispute between high-technology and low-technology designers of solar heating systems is brought into razor-sharp focus by a recent agonizing pronouncement by Marvin Shapiro of Montreal, Canada.* He said, in effect:

Judging solar heating systems solely in economic terms is intolerable. The economic merits of a system can be changed overnight by whim of the legislature—which may reduce house assessments when solar heating is added, or may provide low-interest loans, or provide tax-free status. Such measures change the heat-output-per-buck without changing the heat output itself. In comparing two different solar heating systems, the designer should, of course, take into account their respective economics. But he should give equal attention to a comparison of technical performances. *Technical facts are steadfast!* We need two figures-of-merit and a (steadfast) technical figure-of-merit.**

This view may appeal to some people. But it seems to me to be wrong.

It is wrong because the homeowner's central goal in installing solar heating is to save money. Many solar heating systems cost too much to make economic sense. A few are cheap and reliable and really do make economic sense. The homeowner wants to know which system is the most economic one.

Technical excellence (for example, high efficiency of energy collection) is *not* of primary importance. Nor is it of secondary importance. In fact, it is of no importance at all, relative to the importance of collecting much heat per dollar of overall cost. Efficiency per se is irrelevant. For example, the most efficient solar heating system could be the least desirable from the standpoint of Btu's per buck. (Analogy: The most technically advanced automobile (Rolls Royce?) is far inferior, economically, to a VW. Most people, when deciding which car to buy, give the Rolls Royce—and its technical preeminence—no consideration at all.)

No, Virginia, there is no Santa Claus to give prizes for technical elegance in a solar heating system. High efficiency may have strong esthetic appeal; and efficiency ratings may be as steadfast as Gibraltar. But their pocketbook appeal is nil. And solar heating is, par excellence, a pocketbook matter. A Btu from a solar

*Marvin Shapiro is at the Centre for Building Studies, Concordia University, 1455 Maisonneuve Blvd. W., Montreal, P. Q., H3G 1M8, Canada.
**Quoted (or paraphrased) with permission.

heating system is no different from a Btu from an oil furnace. The key difference is the lower cost (more Btu's per buck) that some solar heating systems provide.

Actually, the steadfastness of technical performance data is not an asset: It is a liability—a fatal defect. Throughout our lives, we shift our habits of buying foods, shoes, cars, etc., as the various prices shift. Much of life consists of adjusting to changing prices. When the legislatures or industrialists change the economics of energy supply, designers of solar heating systems much go back to their drawing-boards and desks and review once again the various design alternatives and decide afresh which system is most cost-effective.

Technical data are steadfast? Yes. The dinosaur and dodo were steadfast—but where are they now?

256

HOW SHOULD WE DEFINE AND MEASURE THE CARRYTHROUGH OF A PASSIVELY SOLAR-HEATED HOUSE?

SUMMARY

I propose that *carrythrough of a passively solar-heated house* be defined as the time it takes, during an overcast period with outdoor temperature 32°F, for the main rooms to cool from 70°F to 60°F, assuming (a) the furnace is left off, (b) the occupants continue to use electric lights, etc., in normal manner, and (c) immediately prior to the start of the cool-down period the main rooms have been kept at 70°F for a long time.

Using this definition, occupants of a passively solar-heated house can easily determine its carrythrough. Also they can evaluate the effect of design changes intended to increase the carrythrough. In addition, designers can compare the carrythrough values of various house and decide which designs of floors, walls, etc., are most cost-effective. The subject comes alive!

INTRODUCTION

A passively solar-heated house that has massive floors and walls will stay fairly warm for a while even on cold nights with the furnace off.

How long will it stay fairly warm? In other words, what is the carrythrough?

Before this question can be answered, a clear and acceptable definition of carrythrough is needed. There has been no such definition.

Is a definition really needed? Yes. The main reason for making floors and walls massive is to provide long carrythrough. Yet if we have no definition of this quantity, we cannot measure it: we cannot evaluate the performance achieved. The merit of any given floor-and-wall system cannot be stated. Different systems cannot be compared. Cost *vs.* benefit analyses cannot be made. Designers lack stimulus to find low-cost ways of increasing the carrythrough.

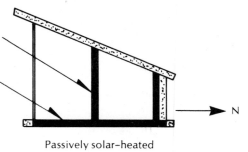

Passively solar-heated
house with massive
floor and walls

BAD DEFINITION

One could define carrythrough in terms of (a) the mass and specific heat of floors, walls, etc., and (b) a 20–F–deg. cool-down range, i.e., from 80°F to 60°F. If, for some particular house, the product of these is 1,000,000 Btu, and if the house needs (from its furnace and/or solar collection system) 500,000 Btu per overcast midwinter day, the carrythrough could be defined as:

$$\frac{1,000,000 \text{ Btu}}{500,000 \text{ Btu/day}} = 2 \text{ days}$$

But such a definition would have little connection with convenience or reality. For these two reasons:

1. There may be no convenient way of producing the postulated starting conditions, for example, no convenient way of bringing all regions of the floor slab to 80°F at the outset of the overcast period. To bring the top of the slab to 80°F is simple enough. But unless the rooms have been kept very hot for a very long time, the lowest regions of the slab are likely to be considerably colder than 80°F. In general, to bring all portions of the slab to a single temperature is very difficult; inasmuch as the slab is made of a material that has high specific heat and high thermal resistivity. The thermal relaxation times are very long—of the order of days or (if the slab is very thick) weeks.

2. A sizable fraction of the heat in the floors and walls may fail to come into play by the time the room temperature has dropped to 60°F. The lowest regions of the slab may still be at, say, 65°F or 70°F when the uppermost regions are at about 60°F. We can have little use for a definition that fails to take into account how fast (or slowly) the heat travels from deep within the slab. (Suppose the "slab" consisted of ledge rock and was one mile thick. What then?)

The thicker the floors and walls, the more unrealistic this definition becomes. Some designers have used this kind of definition, and I would expect that the resulting values of carrythrough are, in some cases, highly misleading. One can imagine changes in design of floors, say, such that, according to this definition, carrythrough would be increased, whereas in fact it would be decreased.

If we wish to put the concept of carrythrough to serious use, we had better be sure that the definition is wisely phrased.

GOOD DEFINITION

Carrythrough is to be defined in terms of a meaningful test. The test is to be carried out in an overcast period in which the outdoor temperature is close to 32°F. (If it is not close, a correction factor can be applied.) In the period just prior to the test the main rooms in the house are to have been kept at about 70°F for a long time—

by any appropriate means such as use of solar collectors, furnace, wood–stove, or electric heaters. By *long time* I mean a time long compared to the carrythrough in question. Note that the starting conditions are easy to arrange, the occupants are subjected to no significant discomfort, and no elaborate set of temperature measurements is needed—there is no need to drill holes in the floor and walls and insert thermometers in these holes.

At a certain suitable starting time, the person conducting the tests reads thermometers placed at various heights in the main rooms and turns off furnaces, wood–stoves, and electric heaters. He cautions the other occupants to continue their normal activities. Hour after hour he reads the thermometers. At a certain instant he finds that the average reading is exactly 60°F. He records this instant. He then calculates the elapsed time. This is the carrythrough.

OTHER CHOICES OF OUTDOOR TEMPERATURE

In practice, the occupants may make such determinations for several outdoor temperatures, such as 14°F, 32°F, and 50°F (i.e., temperatures 18 F degrees apart; the temperatures correspond to −10, 0 and +10 Celsius). From the resulting data a curve of carrythrough *vs.* outdoor temperature can be drawn and correction factors for tests made at other temperatures can be figured out. Persons in cold regions may amass much data on carrythrough with outdoor temperature near 14°F, and persons in warm regions may amass data referring to 50°F. Thanks to the availability of complete curves for a few houses at least, the results from cold and warm regions can be "normalized" and meaningful comparisons can be made.

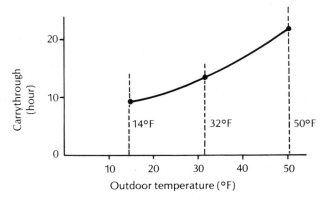

Carrythrough of the Shurcliff house in Cambridge, Mass. This is a big, old, conventionally heated house that has recently been well insulated. The data are based on rough estimates only.

CAN CARRYTHROUGH BE INCREASED EASILY?

Using the above-proposed definition and using the experimental data that may soon become available, designers can start figuring:

What configuration of thermal mass produces the greatest carry-through? Which configuration gives the greatest carrythrough per dollar of capital cost? What minor modifications would increase the carrythrough significantly while increasing the cost only insignificantly? A spurt in hard thought and inventiveness is to be expected.

Here are some possible ways in which (even without increasing the mass of floors, walls, etc.) designers can increase the carry-through:

- Use materials that have greater thermal conductivity, so that the heat can flow faster and farther into the floors and walls and can later emerge from them faster. Different types of concrete and different types of stone have different thermal conductivity.

- Provide larger surface areas of walls. Use deeply grooved walls or perforated walls. Provide channels for circulation of air.

- Use small fans to force air to circulate over the surfaces of floors and walls, or through channels therein. Or merely make provision for encouraging gravity-convection air-flow.

- Install unobtrusive curtains close to the big south windows, for example, orange curtains of gauze-like texture. These will absorb, say, 30% of the incident solar radiation, and air in contact with the curtains will be warmed and will rise toward the ceiling. Collect this local concentration of warm (80°F?) air with the aid of a duct and a low-power blower and deliver it to cold north rooms, cold basement rooms, or to channels beneath the massive floors.

- Leave large voids in the masonry floors and walls and, in these voids, install water-filled drums or mattresses. Instead of using a simple very thick concrete floor, use a thin concrete floor that forms the top of a large-area very shallow, water-filled tank. (Spencer Dickinson did this in a solar house he built in Jamestown, RI, in 1975. The concrete is 4 in. thick and the tank is 16 in. thick. The tank contains water and some stones—the latter being a very cheap means of holding up the concrete while it is being poured and afterwards.)

Of course, attention should be given also to reducing heat loss by improving the insulation of the house and installing thermal curtains on the big windows at night. This may be the cheapest way of increasing carrythrough.

In general, the subject of maximizing carrythrough should be considered carefully by the architect at the time the house is being designed.

FALLACY CONCERNING BENDING A COLLECTOR TO INCREASE THE AMOUNT OF ENERGY COLLECTED

Several inventors have claimed that an ordinary flatplate collector would collect more energy if it were bent along the vertical center-line—bent about 60° so that one half faced almost southeast and the other almost southwest. (See first two figures below.) They claim that one half of the collector would receive an especially large amount of solar radiation in the morning and the other half would receive much radiation in the afternoon. Thus the total amount of energy collected during the day as a whole would be increased, they say.

To disprove this, one may propose two changes:

Change A: Separate the two halves, as in the third figure.

Change B: Turn each half so that it faces straight south, as in the fourth figure.

Clearly, Change A has no affect on the performance. Equally clearly, Change B *improves* the performance—because a single flat collector performs better (for the day as a whole) when facing straight south than when facing far from south.

Yet the overall effect of these changes is to restore practically the initial arrangement shown in the first of the accompanying figures. Which indicates that bending the panel hurts, not helps.

N

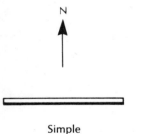
Simple
flat panel facing
south

Bent
panel

Same, but
with the halves
separated

Here each
half has been
turned to face
south

FALLACY CONCERNING THE USE OF A PRISM TO REDUCE THE WIDTH OF A BEAM OF AIR

Some persons have claimed that a prism can be uniquely useful in reducing the width of a beam of solar radiation in air. More radiation, they say, can be directed at an absorber of given area. The accompanying figures show how such persons would use a prism.

The first figure shows how a simple prism can reduce the width of an idealized, 4-in.-wide beam to 2 inches. The incident beam strikes the prism along the normal, and the emerging beam, refracted sharply downward, has a reduced width.

The second figure shows a scheme employing a prism and, beyond it, a slender cylindrical lens that concentrates the radiation on a long slender tube.

Under special circumstances, i.e., within certain narrow limits, these claims are valid. But under a wide variety of practical conditions, they are not.

Persons making the claims may have overlooked some of these facts:

A. Solar radiation includes many components: blue, green, yellow, red, etc., characterized by different wavelengths, and components of different wavelength are refracted through different angles. (See the accompanying figure.) Especially if the prism is oriented so as to produce a large reduction in beam width, the spread in directions of the different components is large. In extreme cases, a very slender flat–plate collector placed so as to receive the red component may be entirely bypassed by the blue component. In summary, the prism performs a service in reducing beam width but a disservice in dispersing the different wavelength–components in different directions.

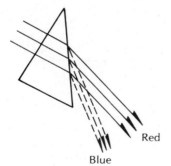

B. Every incident beam includes a small range of directions (has some angular spread) and the prism greatly increases the spread. (Here I am using the most favorable case of a beam consisting of a single wavelength component; if there were a large range of wavelenths, the spread would be greater yet.) Especially if the prism is oriented so as to greatly reduce beam width, the increase in angular spread is great. (There is a famous law in optics that any change in linear width of a beam is accompanied by an equal and opposite change in angular width. This applies to prisms, lenses, mirrors, and all other passive optical devices.) In summary, the prism performs a service in reducing the local linear width of the beam but a disservice in increasing the angular spread. The accompanying figure shows two incident rays that have slightly different directions.

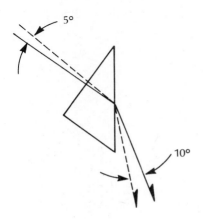

What is the overall conclusion? It is, I think, that prisms should not be used to produce concentrated emerging beams. The increases in angular spread and in spread with respect to wavelength can outweigh the decrease in local width.

If a designer really wants to concentrate a beam, he can do better, ordinarily, by using a lens or a mirror. A fresnel lens, besides being able to concentrate radiation by a factor of 10 or 20 or more, is lightweight and inexpensive. Incidentally, a fresnel lens combines the capabilities of prism and lens while avoiding their weight and cost. (See the accompanying figure illustrating the use of a fresnel lens.) Several firms are routinely producing fresnel lenses 1 to 3 ft. wide and 10 ft. or more in length. With such lenses available, I can think of few worthwhile applications of prisms. Even with one or two cheap flat reflectors, a designer can get excellent results. Crudely curved reflectors can produce even greater concentration. (See the last two figures.)

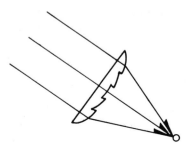

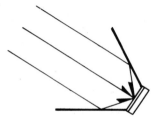

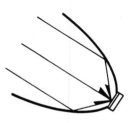

REASONS FOR NOT SEEKING A HIGH DEGREE OF THERMAL STRATIFICATION WITHIN A THERMAL STORAGE SYSTEM

SUMMARY

It is unwise to seek a high degree of thermal stratification within the thermal storage system of a solar heated house. The drawbacks exceed the benefits. The main drawback is a decrease in the overall amount of energy collected during the winter as a whole; this applies whether the storage medium is water or stones. When stones are used there is an added drawback: under certain circumstances the stratification becomes inverted: it works against you—does more harm than good.

However, there is always some temperature rise in the coolant flowing through the collector, and this rise (which the designer should try to keep small) should be put to use: such stratification as it provides should be cherished. What is to be avoided is deliberately increasing the rise in order to increase the stratification.

REASONS APPLICABLE TO A WATER-TYPE STORAGE SYSTEM

INTRODUCTION

Consider a house that has a water-type collector and water-type storage system. Assume that, initially, the water in the tank is cold. If hot water from the collector is now delivered to the top region of the tank, that region soon becomes hot; it does so despite the fact that the bottom region is cold. Thus some thermal stratification is achieved.

It is easy to increase the degree of stratification: merely circulate the water through the collector more slowly so that the outlet temperature is higher; make the tank taller and thinner so as to decrease the rate at which heat from the top of the tank can move (by conduction) to the bottom; install some foam-glass insulating plates within the tank at various heights; or replace the single tank with a series of small tanks (which may be one-above-another or side-by-side).

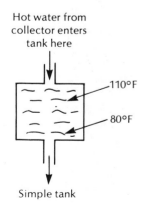

Hot water from collector enters tank here

110°F

80°F

Simple tank

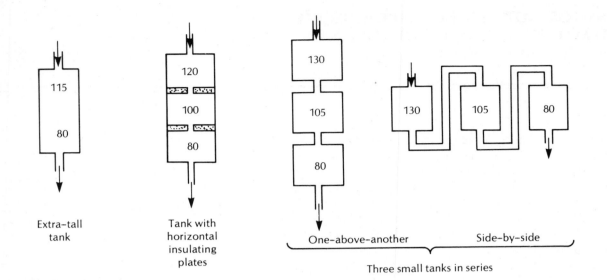

Extra-tall tank

Tank with horizontal insulating plates

One-above-another | Side-by-side

Three small tanks in series

If, in such a storage system, a high degree of thermal stratification is achieved, the designer may be greatly pleased. Designer Jones might say, "Look how successful it is. The top of the tank is very hot, and by drawing water from the top and circulating it through the room radiators, I can keep the rooms warm even though the outdoor temperature is 0°F. I can do this even if 80% of the water in the tank is cold. Also, if the water I circulate to the collector is taken from the bottom of the tank, the collector coolant is so cool that heat-loss from the collector is small and collection efficiency is high. I have the best of both worlds. And the stratification costs me nothing: I merely made a clever design and nature does the rest. Hoorah for thermal stratification!"

My impression is that most designers share Jones' view.

I did too—initially. But I now believe that seeking a high degree of stratification is not desirable. The drawbacks outweigh the benefits. There is one big drawback and several little ones.

BIG DRAWBACK

The big drawback is that with a high degree of stratification less energy is collected than when there is little stratification. This is because, with much stratification, the collector outlet temperature is higher, the average temperature along the collector is higher, and therefore the overall heat-loss of the collector is greater. This would apply, for example, if the temperatures at top, middle, and bottom of the highly stratified tank were 180°, 100°, and 80°F and the corresponding temperatures in a slightly stratified tank were 121°, 120°, and 119°F. The average temperature of the collector served by the former is $(80 + 180)/2 = 130$°F, while the average temperature of the collector served by the latter is $(121 + 119)/2 = 120$°F. Accordingly, the former collector loses much more heat to the outdoors than the latter collector does—despite the fact that the total amount of heat in storage is the same in each case. Different outcomes apply if different temperature distributions within the highly stratified tank are assumed. In some cases the comparison is even more extreme than in the case described above; in some other cases the comparison is less extreme. In a few cases the outcome may be reversed: If, for example, the upper 90% of the tank is very hot and only the bottom 10% is cold, the collector will lose less heat than if the water in the tank had been stirred and made uniform in temperature, but in midwinter such cases are rare indeed. The opposite type of case is far more common.

The typical result is that, with much stratification, the average temperature along the collector is greater, collector losses are greater, and less heat is delivered to the storage tank.

Jones' mistake was in thinking mainly about how low the collector inlet temperature was, instead of how high the outlet temperature was and how high the average temperature in the collector was.

MINOR DRAWBACKS

If the tank is highly stratified, start-up of collector operation must be postponed until later in the morning—otherwise the water delivered to the top of the storage tank will be somewhat cooler than the water that is already there and the degree of stratification will be reduced. Likewise at the end of the day the operation of the collector must be terminated sooner. And on certain slightly overcast days the collector must be left inactive—although energy could be delivered to the tank if the tank-content had been thoroughly stirred. A storage tank designed to accommodate a high degree of stratification is a little more expensive to build and has slightly greater heat-losses unless provided with better insulation. In any event, the stratification is steadily being degraded—wasted away—by the short-circuiting effect of the steel wall of the tank and also by the water itself, both of which have significant thermal conductance and tend to gradually decrease the temperature difference between top and bottom of tank. Of course, the turbulence produced by the water that enters the tank also tends to reduce the stratification.

It is also possible to dampen Jones' enthusiasm about how the distribution of heat to the rooms is benefited by the thermal stratification. For example, Jones may say, "Friday was a cold day, but the top of the tank was hot—hot enough to keep the rooms warm. If I had stirred the water in the tank to bring it all to the same temperature, it would have been too cold to keep the rooms warm." But there is another side to this story. In fact, if the tank had at all times been kept stirred, it would not have gotten so cold; use of the tank would have discontinued sooner, leaving the tank warmer (and auxiliary heat would have been used). And on the subsequent sunny day the average temperature of the tank would have risen higher. A detailed argument would take up too much space, but no detailed argument is needed, inasmuch as the reader already recognizes that, with a high degree of stratification, collector losses are higher and the total amount of heat delivered to the storage system is less. This inevitably means that the amount of solar energy that is ultimately delivered to the rooms is less.

How much stratification should the designer seek or accept? Perhaps he should seek none; I know of no reason for seeking any. But he should willingly *accept* whatever degree of stratification "falls into his lap" as a result of the limitation on rate of flow of coolant through the collector. If the hydraulic resistance of the set of pipes is so high and the power of the water-pump so low that only X gallons of water circulate per minute and, accordingly, the temperature of the coolant in the collector rises (on a sunny day) 30F degrees, the designer should plan on accepting and cherishing about a 30-F-degree temperature difference between top and bottom of the tank. In summary, the designer should not seek stratification but should accept whatever stratification comes naturally.

A final question is: How much extra expense should he go to, in designing another solar heating system, to reduce the resistance of the pipe system and increase the pump power—in order to reduce temperature rise and reduce the drawbacks associated therewith? I do not know how to answer this question.

REASONS APPLICABLE TO A BIN-OF-STONES STORAGE SYSTEM

Here most of the arguments are the same as for the water-type system, and most of the conclusions are the same. The dominant argument is that a high degree of thermal stratification in the bin-of-stones correlates with big temperature rise in the air circulated through the collector and with big heat-loss by the collector.

Again it is true that some temperature rise in the collector is unavoidable, and the stratification that naturally results on this account should be accepted and cherished. Again I wonder how great an effort the designer should make to keep the temperature rise in the collector small.

Unique Drawback

There is a unique drawback to stratification in a bin-of-stones: The stratification can be inverted and can then do more harm than good. The stratification becomes inverted if the circulation of air from collector to top of bin continues for a few sunny hours (causing the top region of stones to become very hot) and then continues during a period of moderate overcast. During this latter period, the coninuing flow of (luke warm) air transfers heat from the top of the bin to, say, the middle, which now becomes the hottest part—and the top of the bin is only luke warm. Thus the main fraction of the heat, temporarily trapped near the middle of the bin, is temporarily unavailable for heating the rooms. (If the hottest water in a tall thin tank were at the middle of the tank, this water would promptly rise to the top; the situation would be corrected within a minute or two. But stones are immobile, and an inverted temperature distribution can persist for a long time.)

TWO BOOKS ON RECENT PATENTS ON SOLAR ENERGY DEVICES

Solar Heating and Cooling: Recent Advances by J. K. Paul, published by Noyes Data Corp., Mill Rd., Park Ridge, NJ 07656, 1977, hard cover. 490 pages, $48.

1977 Solar Energy Inventions and Design Patents by Stanley Garil, PO Box 50003, F street Station NW, Washington, DC 20004, 1978, paperback, 66 pages, $10.

Both of these books are successful. Both present broad surveys of the recent patent literature on solar collectors, storage systems, etc. Both display, in text and drawings, a wide variety of inventions that engineers and inventors will find stimulating.

It is not surprising that both books are frustrating—because patents in general are frustrating. The patent lawyers who write the patents take great pride in using vague words, long circumlocutions, etc., so that the scope of the patent will be as broad as possible and so that the invention will sound new and brilliant. Never do they pin-point the heart of the invention. Never do they explain exactly what is new and what is old. Seldom do they concern themselves with practical or economical applications. Innumerable complications and alternatives are introduced. If the so-called new and ingenious invention is in fact old and puerile, they try to cover up this fact. The persons preparing the drawings are forced to adhere to an archaic style of drawing. The drawings may have been photoreduced to such an extent that they are almost unintelligible. But the reader learns to find his way through the verbal fog and, on the whole, is richly rewarded.

The book by J. K. Paul is clearly the better of the two. It presents long extracts from about 180 patents from the period 1970-1977. Some helpful introductory paragraphs are included also. There is a table of contents, an author index, and a patent-number index. The typography and drawings are clear.

But the book by Garil runs a close second, having two points of superiority: It covers more patents (about 300), and nearly all of them are very new—1976 and 1977. Unfortunately the material is crudely photoreproduced and there is no table of contents or index. It costs only $10, as compared to $48 for the book by Paul.

A PERIODICAL ON THE SAME SUBJECT

A periodical entitled *Solar Energy Patents* is published by Impact, PO Box 1972, Estes Park, CO 80517. It is a monthly, edited by A. L. Anderson, and costs $40 per year.

NOTES ON SOME VERY OLD PATENTS ON
SOLAR HEATING

R. W. Bliss assembled, many years ago, a three-volume set of old patents on solar heating. He kindly lent me this set. Notes on some of the most interesting of these patents are presented below. For each I indicate the US patent number, the date of issuance, and the inventor.

705,350   7/22/1902   P. G. Hubert

Water-type collector employing two sloping plates that are so close together that water flowing downward between them makes contact with both and picks up heat from the irradiated (upper) sheet.

(A few years ago S. C. Baer and Ronald Shore made collectors employing two corrugated sheets of aluminum that nest close together with water flowing between them. One of H. E. Thomason's patented inventions involves a corrugated sheet that is folded so as to constitute upper and lower nesting portions with water flowing between.)

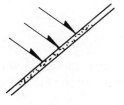

761,596   5/31/1904   Eric Moss

A large cone, or funnel, receives solar radiation and reflects it toward the axis of the cone. Here a small black coil containing flowing coolant receives the radiation. The assembly is continually reoriented (re-aimed) so as to track the sun.

(A few years ago W. G. Steward and/or J. F. Kreider invented a collector employing a portion of a sphere, rather than a cone, and a long slender coil the axis of which passes through the center of the sphere and aims at the sun. The portion of sphere remains fixed and the coil itself is continually swung (about a point corresponding to the center of the sphere) so as to track the sun.)

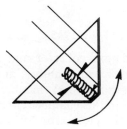

842,658   1/29/1907   C. L. Haskell

Along the upper edge of the sloping, roof-mounted collector box there is a long horizontal tube in which there are many small, equi-spaced holes that feed water to the box in many small streams.

937,013   10/12/1909   M. L. Severy

A transparent structure with circular or linear refracting elements focuses radiation onto a small receiver.

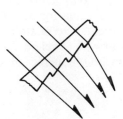

966,070   8/2/1910   W. J. Bailey

Collector employing a set of slender, widely spaced tubes that are affixed to black sheets of metal. The absorbing area is large and the thermal capacity of the system is small.

980,505   1/3/1911   W. L. R. Emmet

The collector tubes are vacuum jacketed as by means of glass tubes that include plain areas that admit radiation and silvered areas that reflect radiation toward the collector proper. Large cylindrically curved mirrors are employed to funnel radiation toward the tubes-and-jackets.

(Recently several manufacturers have produced vacuum-jacketed tubular collectors and many have employed cylindrically curved mirrors.)

1,042,418   10/29/1912   A. H. Evans

The collector includes two oppositely undulating metallic sheets, joined together, with water flowing between them.

1,801,710   4/21/1931   C. G. Abbot

Collector employing glazing consisting of close-packed row of glass tubes that are slightly elliptical in cross section. The tubes are evacuated and there are sealing strips between them.

2,388,940   11/13/1945   R. H. Taylor

Combination collector-and-storage system with thin front chamber, thick rear chamber, and an insulating septum between the two. The thin (south) chamber receives solar radiation, and the heated water in this chamber circulates (by gravity convection, via slots in the top and bottom of the septum) to the thick chamber, which stores much heat.

(The system has some similarity to the Shawn Buckley system and to my system S-48.)

2,484,127   10/11/1949   W. Stelzer

Air-type, gravity-convection, solar-radiation-collecting-and-storing system that includes a valve that automatically closes at night to prevent reverse flow. The valve, of near-frictionless type, consists of a lightweight metallic vane which, when closed, makes contact with the surface of liquid in a shallow tray.

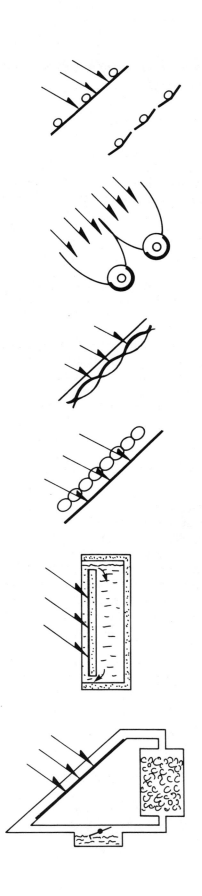

2,553,073   5/15/1951   R. E. Barnett

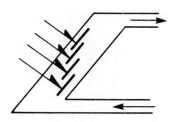

Air-type collector panel containing, as absorber, a set of spaced, partly transparent and partly absorbing, overlapping plates.

(A somewhat similar set of absorbing plates was installed in the collector mounted atop the house built by G. O. G. Löf in Denver in 1957.)

2,594,232   4/22/1952   C. L. Stockstill

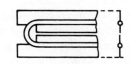

Absorber consisting of an extruded combination: aluminum tube with integral aluminum fin. If suitable cut-outs are made, the extrusion can be bent into serpentine pattern.

WHIPPING BOY: THE GOVERNMENT

Most people blame the government for the slow progress on solar heating. They forget that the main purpose of the government is merely to govern. They expect it to act also as a kindly father, smiling mother, and a rich uncle. They expect it somehow to turn its hand to inventing, developing, demonstrating. They expect it to encourage, stimulate, and coordinate. They expect it to issue wise guidelines and regulations. They expect it to provide subsidies big enough to make overly expensive systems look like bargains.

Examined in detail the views of our citizens on these subjects are highly contradictory, i.e., largely self-canceling. See, for example, DOE's 40-page "Consumer Briefing Summary # 7" of October 1978. It summarizes what hundreds of complaining people said at various regional meetings held in the summer of 1978. Nearly all of these people criticized the government for not going far enough and others criticized it for going too far.

A sobering fact is that, ordinarily, a government cannot invent, or manufacture, or establish prices, or guarantee, or advertize, or sell, or deliver, or install, or repair.

What, then, can it do? It can alarm and confuse. It can frighten inventors and producers into inaction. Its powers are enormous: they include the right to regulate or prohibit, the right to stimulate favored entrepreneurs by injecting large amounts of money, and the right to direct a searchlight of publicity on anything it likes or dislikes. The mere hint that some of these powers may be used is often enough to frighten private groups into inaction or retreat.

But before anyone laughs too loudly at the efforts of DOE, HUD, and other agencies, he should remember these handicapping realities:

Dozens of governmental agencies are involved, and their responsibilities overlap. The regions of overlap are large, are not well understood, and form a kind of no-man's-land.

The agencies' roles are limited by congressional bills. Certain helpful things that a given agency could do are forbidden. Certain foolish things are made mandatory.

The agencies cannot act fast. Fast actions are usually arbitraty actions. Government agencies are forbidden to be arbitrary. They must be fair to all—which entails long delays.

They cannot display courage. People who display courage occasionally make big mistakes. If DOE make big mistakes, Congress may cut its funds.

They operate in a continual state of seige: They are beseiged by a horde of citizens asking for pamphlets, interviews, speakers, reports, grants, contracts, and subsidies, and by congressmen trying to get favored treatment for constituents.

The government's efforts certainly are laughable. But could you and I do any better?

GOVERNMENT STANDARDS ON SOLAR HEATING EQUIPMENT: ARE THEY FOCUSED ON THE WRONG THING?

It seems to me that the men who are writing government standards on solar heating equipment have been concentrating on the wrong thing. Most of the standards they draw up deal with durability and efficiency, instead of with cost-effectiveness.

I remember the old golfing joke: Smith says, "Your ball went into the pond; why are you looking for it on the fairway?" Jones replies, "I am trying to be practical. Searching the pond is difficult and messy. But searching the fairway is fast and straightforward."

What the potential buyer of solar-heating equipment really wants is equipment that is cost-effective. He wants to know which system, over a 20-year period, will deliver the most heat at the smallest overall cost.

Do the standards writers prepare standards on this? In comparing the various makes of equipment, do they list values of "Btu's per buck"?

They do not.

Why not? If asked, they would reply, I believe, along these lines: "The subject is too difficult and messy. No one can predict the number of Btu's delivered, because it depends on so many factors, such as site, climate, what type and size of storage system is used, and how the residents operate the system. All of these factors are outside our control. Cost, also, is difficult and messy; suppliers keep changing their prices; installation costs are hard to predict; maintenance costs are unknown."

Then why don't the standards writers give up? If they can't answer the crucial question, why don't they remain silent (or stick to issues of safety, which everyone knows to be important)?

Here their reply might take this form: "We are trying to do our best. Trying to be helpful. So we write standards on collection efficiency and durability. We think it will be helpful to buyers to know which equipments have high collection efficiency and high durability."

But here the tragedy of standards reaches its climax. The fact is that the most efficient and most durable equipment may or may not be the most cost-effective. Conceivably it may be the least cost-effective—the worst buy. A Rolls Royce may be the most efficient and durable automobile, but its cost-effectiveness is far below that of a Toyota. A $15 pen may perform superbly, but it is less cost-effective that a typical 50c pen.

To me it is frightening to see a government agency (for example, the agency recently set up by the State of California) set up standards on efficiency and durability and give the public vast amounts of information on these topics, despite the fact that the correlation between these topics and cost-effectiveness is dubious and may even be negative. Is such information really helpful? Or does it distract people from what is truly important: cost-effectiveness?

And won't manufacturers too be distracted? Will they not be tempted to modify their designs so as to increase efficiency and durability even at the risk of decreasing cost-effectiveness? This is what I fear.

HOW CAN THE GOVERNMENT WRITE STANDARDS ON SOLAR HEATING SYSTEMS THAT CONSIST MAINLY OF CLEVER DESIGN?

Some of the most cost-effective solar-heated houses are those employing passive solar heating, i.e., those relying on clever design of the house itself. The house itself is the collector and the storage system. Examples of such houses are: Hunt House, Bier House, Saunders House. Equally clever designs are used in the Cambridge School, in Benedictine Monastery, and dozens of other passively-solar-heated buildings.

The variety of passive designs is necessarily great; probably it is comparable to the variety of the buildings themselves. This variety is great indeed, as is apparent when one considers

The variety of building sites:
city or country, open plain or forest, valley or hilltop, warm region or cold region, clear region or cloudy region

The variety of types of building:
residence, apartment building, office building, school, church, factory

The variety of sizes and shapes:
500 to 500,000 square feet, one-story or multi-story, with or without basement, square, oblong or L-shaped

The variety of materials used:
wood, brick, concrete, stone, adobe, steel

The variety of styles:
Cape Cod, modern, etc.

The variety of elegance:
high-tech. professional or do-it-yourself

Combining all these choices, one finds that there are literally hundreds of classes of buildings—implying also a very large number of designs of passive solar heating systems.

THE PROBLEM

The problem is that the men who set out to draft standards on passive solar heating systems will find the task too big, too difficult. To define several hundred classes of buildings and passive solar heating systems, and to draw up standards for them, would be an enormous undertaking, and the results might fill more than 1000 pages. If the men attempt this task, they are likely to fail. Even if they were to succeed, who could wade through the 1000 pages? And would not the subsequent additions, changes, clarifications, etc., be endless?

Does it matter if they fail? Does it matter if no standards are written for, say, 50 promising kinds of passive solar heating systems? Yes, because if there are no standards, there will be no tax abatement offered. And because no tax abatement is offered, buyers will turn away. Some will turn to (less cost-effective) active systems, and some will spurn solar entirely and continue to burn oil.

Failure to provide standards for passive systems (while providing standards and tax abatement for active systems) will be a crippling blow to passive solar heating.

WHY IS THE PROBLEM AGONIZING?

It is agonizing because it so strongly penalizes the very best systems: the most durable and most cost-effective systems. It gives a boost to active systems, which so often are very expensive and complicated, and inhibits passive systems, which are superior in so many ways. The solar heating industry as a whole will be diverted from the most promising avenue.

CONVENIENCE—AN INSIDIOUS CONCEPT

To write standards on ordinary types of flat-plate collectors is highly convenient: The panels are already available, transporting them by truck is easy, terminologies and test methods are available, testing laboratories already exist. Thus it is logical that the standards writers deal first with these panels.

But there is nothing convenient about writing standards for passive system. Being integral with a house, a passive system is hard to define and demarcate. No two systems are alike. There are no big mass-production factories. Passive systems cannot be shipped by truck. Terminologies and test methods have not been established. Standards writers will probably tackle passive systems last—and may *never* get to the cleverest, strangest, and most cost-effective systems.

The result is that, on grounds of *convenience* to standards writers, active systems will be taken up promptly: standards will be prepared, and tax abatements will be allowed. But passive systems will be left in the lurch.

A hypothetical analogy may be helpful. Suppose that the mayor of Venice worries about his city's transportation system. He resolves to improve it—with standards and tax incentives. Will he deal first with boats or cars? He chooses cars, because they are more standardized, easier to specify, easier to test. He soon issues standards on cars and begins allowing tax abatements. But he finds that boats are so highly variable, hard to talk about, hard to test, the he turns his back on them. What is the consequence? Use of cars increases, and use of boats declines. The few streets become hopelessly jammed, while the canals remain almost empty. Because the mayor fell victim to convenience, the city's transportation system has been made worse.

PROTECTING CONSTRUCTIVE RIVALRIES

Within the economy of our country there are countless rivalries: *constructive* rivalries. Some are among different industries, some are among different branches of the same industry. Well-meant interference by the government can pervert such a rivalry and do far-reaching harm. The work-a-day, free-market economic forces, if left alone, often accomplish wonders. Government intervention can spoil this.

SUGGESTION

I suggest that the government drop all financial inducements in this field: provide no inducements for active systems or passive systems. Then the whole problem evaporates. And people who find how to build and install truly cost-effective systems will do so, and those who do not will abstain. Architects and solar designers can forget about 1000-page rule books and high-paid lawyers and can go back to what they are fitted for: building systems that are even more clever, even more cost-effective.

The elimination of government financial inducements will be a blow to manufacturers producing equipment that is not cost-effective. But it will bring the cost-effective systems to the fore, encourage inventors to invent cheaper systems, and prepare the way for an era in which solar heating can stand on its own feet and advance rapidly.

RESPONSE TO A REBUTTAL CONCERNING GOVERNMENT STANDARDS

2/19/78

A prominent solar-heating official recently indicated strong dissent from my views that solar heating is a young industry, government standards will inhibit innovation, and efforts to insure that buyers get their money's worth would entail too much bureaucratic control. His rebuttal included these main arguments:

1. The government must make sure that our citizens are protected from hazards.
2. The blame for the energy crisis lies mainly with industry.
2. For a solution, we must look to government.
4. Solar heating is a mature technology and few important innovations are to be expected.
5. The government must make sure that people do not spend money on dubious products.

My response is:

Regarding Item 1:

The government has no general commitment to protecting us from hazards. Consider its attitude toward the 50,000,000 smokers and the 500,000 annual deaths from cancer. Does the government make any serious effort to stop this? Or consider the 1,000,000 persons each year in skiing accidents. Think of the 500,000 guns and revolvers sold each year.

Regarding Item 2:

Aren't we all to blame—industry, government and the people?

Regarding Item 3:

The Government's record on energy saving is not good. Only a few years ago the government spent $700,000,000 on the development of a supersonic transport plane that would have been the biggest energy-waster of all time, burning one ton of fuel per passenger per trip. Also it took our government two years merely to choose a site for SERI.

Regarding Item 4:

Solar heating is not a mature technology. It is comparable to the automobile technology in 1910.

Regarding Item 5:

The government does not (and cannot and should not) make a systematic effort to prevent people from buying dubious products. Walk through any shopping center and in ten minutes you see enough dubious products to turn your stomach: flashy clothing, worthless drugs and lotions, foods consisting mainly of fats and sugars. And what about cigarettes, whiskey, and books on astrology? What could be more dubious than these? But in any event we would not want to live in a country where the government dictated what we could buy and could not buy. Even the Soviet Union gives its people some freedom to buy goods that are dubious or even harmful. It is a nightmare that our government—which does *not* know how to make truly successful and cost-effective solar-heating systems—is preparing to become dictator in this field.

DEFINITION OF A FACTOR-OF-MERIT FOR USE IN COMPARING DESIGNS OF SOLAR HEATING SYSTEMS FOR TYPICAL HOUSES IN USA

SUMMARY

A factor of merit is defined in detail.

The arbitrary and imprecise nature of the factor is discussed.

The great need for such a factor—despite its obvious limitations—is spelled out.

A work sheet for use in actually computing the factor for a given solar heating system is presented.

INTRODUCTION

It is, of course, impossible to arrive at an objective, fair, accurate, generally acceptable scheme for *quantitatively* comparing various diverse designs of solar heating systems for use in typical houses in USA. Many different considerations—and different *kinds* of considerations—enter. Some are esthetic. Some are engineering. Different persons have different requirements. Some of the necessary engineering data are not available. Some schemes that may be practical in 1990 are not practical today. Cost is crucial, but prices of components are changing fast and unpredictably. The locations and topographical setting of the house may have important bearing on choice of design.

Yet a designer cannot be content with such vague appraisals as "A is somewhat better than B." (It is 10% better, or 1000% better? Would one or two minor improvements rescue B?). What is needed is a quantitative evaluation—even if only a very rough one.

It is intolerable, also, if the designer, each time he wishes to make quantitative comparisons of several designs, must think through afresh all the pertinent considerations and must decide afresh what weights to assign to the considerations. It is intolerable also that he has no ready way of explaining to colleagues what weights he used.

I here propose a comprehensive factor-of-merit, hoping that friends will point out ways of improving it. Helpful comments have already been received from G. F. Tully.

Even if the specific scheme proposed is too rough and premature to be used in formal manner, it may constitute a useful checklist of important considerations and criteria.

ASPECTS EXCLUDED

To simplify matters I have ignored possible extensions of the solar heating system to supplying energy to the domestic hot water system or to cooling the house in summer. Also I have ignored the cost involved in insulating the house walls, roof, etc.

I ignore also the merits of individual components per se. Thus I do not attempt to appraise the collector per se, or the energy store per se, or the control system per se. What really counts is the system as a whole.

I ignore efficiency. The house occupant wants warmth, convenience, economy, and reliability. If all these goals are achieved, he does not care whether they are achieved efficiently or inefficiently.

I ignore problems related to building codes, compliance with federal and state standards on solar heating systems, eligibility for receiving government grants, tax abatements, and low-cost loans.

DETAILS

I call the overall factor-of-merit F and I define it as the product of many constituent factors F_1, F_2, etc. Thus,

$$F = F_1 F_2 F_3 F_4 \dots\dots$$

The constituent factors are arranged in groups. There are three groups:

esthetics:	how the house looks from outside and inside; adequacy of daylight in rooms; adequacy of view outward from the rooms
performance:	ability to keep house warm enough on a typical day in winter; ability to keep house warm throughout a succession of overcast winter days; ability to keep house warm even if electric supply fails for a day; convenience of use; freedom from various hazards, dangers, worries
economics:	low-construction cost; low operating cost; speed of construction; ease of repairing and modifying the system; avoidance of use of strategic materials

I have tried to make the set of constituent considerations an orthogonal one, i.e., a clean one in which each major consideration squarely governs one factor and plays no role in any other factor.

The constituent factors are of normalized type. Specifically, the value of each factor is approximately *unity* for a "typical, good design," and actual values range from slightly above unity to a value well below unity.

Each factor is of affirmative type: A higher value indicates a more desirable solar heating system; A low value indicates a poor system. The value *zero* is assigned to a condition that would render the entire system completely unacceptable.

Use of normalization (typical good performance indicated by unity, and other values ranging from somewhat above or below unity) has several important merits:

> If the designer later introduces additional factors, they have little effect on the overall product (on F, that is) because multiplying a quantity by a number close to unity affects it only slightly. Thus, F-values obtained in the new way can still be compared, at least roughly, with values obtained in the old way. (Some gaps are left in the series of F-symbols to allow room for additional factors.)

> Conversely, if the designer omits a few factors for any reason (for example, if data are lacking and he does not know how to arrive at an approximation) the effect on F is small.

> The value of F itself is somewhere near unity, i.e., is of "managable" size.

ESTHETICS: CONSTITUENT FACTORS

F_1 Outside appearance: appearance of house as judged by a visitor approaching it; appearance of south wall and other walls, and of roof; extent to which trees and shrubs may be retained, and freedom to choose types, heights, and locations of trees and shrubs; some illustrative values of F_1 follow.

1.3 excellent: Solar heating system imposes no limitations.

1.0 fairly good

0.5 very poor: External appearance of house is very poor; no trees or shrubs can be tolerated at south side.

F_2 Inside appearance: extent to which size of windows, and their locations, are normal, permitting normal amount of daylight to enter and permitting occupants to enjoy normal views of the environs

1.3 excellent: The system imposes no limitations.

1.0 fairly good

0.2 very poor: No daylight enters, and there is no view.

F_3 Other esthetic aspects: freedom from noises from blowers, dampers, etc., from smells of chemicals, hot organic materials, fungi, etc., and from other esthetic nuisances

1.2 excellent

1.0 fairly good

0 intolerably bad

PERFORMANCE: CONSTITUENT FACTORS

F_{11} Room temperature: Adequacy of nominal temperature of rooms on typical day in January, assuming that there is no bunching of overcast days, no use of furnace or electric heater, no snow, no failure of electric supply

1.2   excellent: 70°F

1.0   fairly good: 60°

0.7   50°F

0.3   40°F

0.1   30°F

Note: A solar heating system that cannot keep the rooms hotter than 40°F on a typical day in January may nevertheless be of considerable value inasmuch as (a) at such temperature, water in pipes does not freeze, (b) electrical heaters can easily bring the temperature in one or two rooms up to 70°F at moderate cost, (c) in warmer months (March, e.g.) the solar heating system by itself may be able to keep the rooms at 70°F.

F_{12} Carrythrough: (For discussion of this term, see p. 256)

1.4   excellent: 4 days

1.0   fairly good: 2 days

0.5   poor: 1 day

F_{13} (Re snow, etc.) Ability to perform well during or just after a heavy snow or sleet storm: Snow and sleet may blanket the collector window and auxiliary mirrors and may be difficult to remove. Florida residents may omit this term. The weightings proposed below refer to, say, Maine.

1.2   excellent: unaffected by snow and sleet (or removal is immediate and automatic)

1.0   fairly good

0.6   very poor: Small amount of snow or sleet halts collection, and removal of snow or sleet is very difficult.

F_{14} (Re collection with electricity off) Ability to collect energy even when electric supply is off: It may be off because of failure of the electric utility's generator, because of breakage of the supply line or because of rationing regulations.

1.2   excellent: Collection is unaffected.

1.0   fairly good

0.8   none: Collection ceases until the supply is reinstated.

Note: Here we assume that the year is 1985, the energy crisis is severe, and the electric supply will be off eight times per winter for eight daytime hours on each occasion. If the energy crises were much more severe than this, the value *0.8* should be replaced by a much smaller value.

F_{15} (Re Delivery with Electricity Off) Ability to deliver stored energy to the rooms even when electric supply is off for eight hours twice per month:

    1.2   excellent: Delivery is unaffected.

    1.0   fairly good

    0.7   none: Delivery ceases until supply is reinstated.

F_{16} Extent to which the utilization (assignment) of space within the main stories of the house is at the occupant's discretion.

    1.3   excellent: The space is fully at his discretion.

    1.0   fairly good

    0.3   very poor: Half of the space is preempted by solar heating system, and this half includes several key locations.

F_{17} Extent to which the utilization of space within the attic and basement is at occupant's discretion (or extent to which the system requires that there be an attic or basement—which otherwise might be omitted):

    1.2   excellent: The space is fully at his discretion.

    1.0   fairly good

    0.7   very poor: Great majority of such space is preempted by solar heating system.

Note: Preemption of attic and basement space is less serious that preemption of space in main stories. Thus, the values proposed here are not very far from 1.0.

F_{18} Degree to which the system operates without attention by the occupants:

    1.2   excellent: Even if occupants are forgetful, or absent, the system continues to perform well.

    1.0   fairly good

    0.3   very poor: Occupants must keep the system on their minds almost continually and must make adjustments about five times per typical day.

F_{19} Degree to which system itself escapes damage despite unusually extreme conditions of irradiation, low outdoor temperature in winter, high temperature in summer, or heavy snowfall:

    1.2   excellent: Virtually impossible for system to be damaged

    1.0   fairly good

    0.2   very poor: Various quite possible conditions are likely to result in serious damage to system.

Note: Some kinds of storage materials or containers can be damaged by extreme cold. Some kinds of storage materials, containers, sealants, and insulating materials can be damaged by very *high* temperature.

F_{20} Extent to which health of occupants is independent of malfunction or overstressing of solar heating system:

1.1 excellent: full independence

1.0 fairly good

0.2 very poor: Various quite possible situations could result in serious harm to occupants.

Note: We include here hazards from fire—or release of toxic chemicals—attributable to the solar heating system

F_{21} Extent to which the solar heating system is free of threats from *external* agents such as hail, rain, bees, wasps, birds, dogs, children, vandals, acid vapors in atmosphere:

1.2 excellent

1.0 fairly good

0.3 very poor: Such agents might well do much harm.

F_{22} Extent to which the solar heating system is free of threats from *internal* agents such as rats, mice, centipedes, flees, snakes, fungi, mushrooms, algae, children, dogs, cats:

1.2 excellent

1.0 fairly good

0.3 very poor: Such agents might well do much harm.

ECONOMICS: CONSTITUENT FACTORS

F_{31} Extent to which the cost of constructing the solar heating system can be kept low: (Here we *exclude* the cost of inventing the system and the cost of the engineering planning.) If expensive, specialized mass–assembly tools and assembly lines would be required (and would presumably be used in constructing *many* heating systems), we assign to the individual house only a few percent of such large expense. We exclude also the cost of the land and costs associated with insuring that sunlight approaches the collector without obstruction. We exclude the cost of the very–high–performance thermal insulation required in house walls and roof (but we include, of course, the cost of special insulation that is part of the solar heating system). We exclude the cost of any *supplementary* heating system

   1.5   Construction cost is less than 5% of value of house as a whole.

   1.0   Construction cost is 10% of value of house as a whole.

   0.7   Construction cost is 15% of value of house as a whole.

   0.4   Construction cost is 25% of value of house as a whole.

F_{32} Extent to which annual cost of operation and normal servicing can be kept low: (Excluded are occupant's efforts and worry, major unexpected repair work, and cost of oil, electricity, etc. for supplementary heat.)

   1.3   excellent: Annual cost is extremely low, i.e., less than 0.1% value of house.

   1.0   fairly good: 0.3% of value of house.

   0.5   very poor: 1.5% of value of house.

F_{33} Extent of freedom from long delays in construction as a consequence of unique and difficult techniques involved:

   1.3   excellent: No significant difficulties.

   1.0   fairly good

   0.7   very poor: difficulties may cause 1–year delay.

F_{34} Ease of making major repairs: Here we include especially great difficulties connected with repairs of components. Example: replacing broken window panes, including the least accessible panes, repairing or replacing tanks that have developed leaks as consequences of corrosion.

   1.3   excellent: It is almost inconceivable that such difficulties will arise; or repair would be extremely simple.

   1.0   fairly good

   0.5   very poor: Such difficulties may well arise and repair would be very difficult.

F_{35} Extent of freedom from possibilities of serious shortcomings in design or construction: Here we take into account the dire consequences that sometimes result when daring new designs (never fully tested) are put into effect. Costs may soar. Performance may fall far short of expectation.

 1.3 excellent: Design and construction are so straightforward and simple that it is almost inconceivable that serious shortcomings will appear.
 1.0 fairly good
 0.7 very poor: Shortcomings may well show up.

F_{36} Extent of avoidance of use of materials and skills that are in short supply: Here we take account of the national strategic importance (not dollar cost) of such materials and skills.

 1.3 excellent: No strategic materials or skills required.
 1.0 fairly good
 0.5 very poor: Much use is made of such materials and skills.

F_{37} Extent to which system layout, etc., could be changed, from time to time, as the owner's requirements change:

 1.3 excellent: Large changes could be made easily.
 1.0 fairly good
 0.8 very poor: Very difficult to change main features.

Note: A 14-inch-thick, poured concrete, steel-reinforced Trombe wall is very difficult to alter or remove.

F_{41}, F_{42}, etc. All other considerations: Individual designers may introduce special factors to take into account other important criteria, such as:

 helping heat domestic hot water
 helping keep house cool in summer
 providing greenhouse space
 eligibility for grants, tax abatement, etc.

Table 1 is a form, or worksheet, that may be used in computing the factor-of-merit of an actual system. Or it may be used merely as a checklist of the parameters that solar-heating-system designers should keep in mind.

TABLE 1
WORK SHEET FOR COMPUTING FACTOR OF MERIT

		VALUE ASSIGNED

ESTHETICS

F_1 outside appearance
F_2 inside appearance
F_3 other esthetic aspects

PERFORMANCE

F_{11} room temp. achieved
F_{12} carrythrough
F_{13} re snow
F_{14} re collection with electricity off
F_{15} re delivery with electricity off
F_{16} re space in main stories
F_{17} re space in basement and attic
F_{18} re occupant attention needed
F_{19} re proof against damage
F_{20} safety of occupants
F_{21} re external threats
F_{22} re internal threats

ECONOMICS

F_{31} construction cost
F_{32} operating cost
F_{33} delays in construction
F_{34} ease of repair
F_{35} re design errors
F_{36} re stratetic materials, etc.
F_{37} re ease of alteration

OTHER

F_{41} all other considerations   _____

F the overall product

APPENDIX:
U.S. PATENTS ON INVENTIONS BY
W. A. SHURCLIFF

Most of the patents were obtained several decades ago when I was employed by American Cyanamid Co. and by Polaroid Corp. They own most of the patents. The U.S. patent numbers and dates of issuance are:

2,329,657	9/14/43	Spectrophotometric equipment
2,347,066	4/18/44	Spectrophotometric equipment
2,347,067	4/18/44	Spectrophotometric equipment
2,364,825	12/12/44	Spectrophotometric equipment
2,369,317	2/13/45	Color-translating microscope and color-shift photography
2,383,346	8/21/45	Spectrophotometric equipment
2,453,163	11/9/48	Improving X-ray photography
2,705,757	4/5/55	Converting Polaroid camera for use as emergency wartime gamma-radiation dosage meter
2,707,237	4/26/55	Increasing the sensititivy of Polaroid camera gamma-radiation dosage meter
2,730,625	1/10/56	High-precision wartime gamma-radiation dosage meter costing less than $5.
2,747,103	5/22/56	Instant-reading wartime gamma-radiation dosage meter costing less than $2.
2,750,515	6/12/56	Gamma-radiation dosage meter indicating dose to user, not dose to meter
2,779,017	1/22/57	Information processing system for radar PPI screen (with E. H. Land)
2,827,823	3/25/58	Narrow range synchronizer for 3-D movies
2,854,883	10/7/58	Wide range synchronizer for 3-D movies (with R. C. Jones)
2,968,994	1/24/61	Device for focusing a microscope automatically
3,037,423	6/5/62	Device for focusing a slide projector automatically in 0.2 second
3,247,379	4/19/66	Gamma-radiation dosage meter
4,117,882	10/3/78	Thermal storage system employing phase change material

Some of the inventions and patents turned out to be of little or no use. Most turned out to be useful. The most useful was the device for automatically focusing a slide projector; automatically focusing projectors have been mass-produced by several manufacturers and are now in widespread use.

Some of my inventions that proved to be useful but were not patented include a carburator intake silencer for automobiles (1930) and a monochromator (for Raman spectroscopy) that provides doubled intensity without sacrifice of spectral purity (1952).

In 1955 I discovered the first and only known method of distinguishing right-circularly-polarized light from left-circularly-polarized light by means of the naked eye; see *J. Opt. Soc. Am. 45,* 399 (1955). In 1958 I discovered a new phenomenon of color vision: the greenish-yellow blotch effect; see *Nature, 183,* 202 (1959) and *J. Opt. Soc. Am. 49,* 1041 (1959).

Index

292

NOTES

NOTE